AF830616

Heating Systems
Troubleshooting & Repair

Heating Systems Troubleshooting & Repair

Maintenance Tips & Forensic Observations

John Certuse, P.E.

INDUSTRIAL PRESS, INC.

Industrial Press, Inc.
32 Haviland Street, Suite 3
South Norwalk, Connecticut 06854
Phone: 203-956-5593
Toll-Free in USA: 888-528-7852
Fax: 203-354-9391
Email: info@industrialpress.com

Author: John Certuse, P.E.
Title: Heating Systems Troubleshooting & Repair:
Maintenance Tips & Forensic Observations
Library of Congress Control Number: 2019936359

ISBN (print): 978-0-8311-3640-6
ISBN (ePDF): 978-0-8311-9518-2
ISBN (ePUB): 978-0-8311-9519-9
ISBN (eMOBI): 978-0-8311-9520-5

Editorial Director/Publisher: Judy Bass
Copy Editor: Judy Duguid
Compositor: Patricia Wallenburg, TypeWriting

industrialpress.com
ebooks.industrialpress.com

10 9 8 7 6 5 4 3 2 1

CONTENTS

For videos pertaining to subjects and procedures
depicted in this book please visit:
www.CertuseHeatingSystems.com

PREFACE

Most technicians leaving trade school are typically focused on passing a license exam that, for the most part, addresses the installation of new equipment. The fact is, however, that over the course of their career, most of the work they perform is not installing new equipment but, rather, repairing and maintaining older, existing systems. In fact, some technicians may find themselves servicing equipment that is more than 100 years old (Figure P-1).

FIGURE P-1 Early Turn-of-the-Century Boiler Still in Use Today

Heating Systems Troubleshooting & Repair: Maintenance Tips & Forensic Observations was written for those technicians installing new systems as well as for those keeping older systems in proper working order. This book addresses many issues, from boiler and furnace design to maintenance requirements of both the appliance and the entire system heating the home. It also addresses the need to change attitudes toward maintenance.

Prevailing Attitudes

It has been said that we live in a "disposable society" where older appliances are rarely repaired, with service technicians preferring to replace equipment needing more advanced repairs. This attitude extends to home heating appliances that, due to the fault of many, the homeowner included, are often left in service until small repairs are no longer sufficient and the appliance requires replacement.

Then there is the commonly repeated philosophy of "If it ain't broke, don't fix it." Following this philosophy means only addressing repairs *after* equipment has broken down and can result in unexpected consequences.

Where does the responsibility fall for boiler and furnace failures? That responsibility belongs to everyone involved—but to different degrees. On rare occasions, it is the manufacturer who is responsible, but more often it is the homeowners and those employed to maintain these appliances.

The manufacturing and installation of an appliance are onetime occurrences, performed and then done. Required preventative maintenance actions, however, are ongoing and needed throughout the service life of the appliance.

The Importance of Preventative Maintenance

Residential boilers and furnaces are designed to run efficiently and reliably throughout their design life. But to achieve longevity of life and efficient operation, the appliance and system must be maintained through a regimen of preventative maintenance actions throughout the equipment's service life. It is these preventative maintenance actions that determine the quality of the service life of the appliance as well as the likelihood of unexpected equipment breakdown, costly equipment replacement, and building damage.

A Team Effort

This book is intended to familiarize new service technicians of the overall maintenance requirements of a heating system and to help them to realize that maintenance is not limited to one particular component of a system without regard to others.

It's a team effort to keep home heating appliances running safely, efficiently, and reliably. The team includes both the manufacturers of appliances that provide training and replacement components and the local service organizations that install these appliances and perform needed repairs. More importantly, this team includes the homeowner or building owner who must also support the service technician's recommendations by scheduling annual maintenance *allowing* the technician to perform needed maintenance repairs.

In the end, each party needs the other to make sure a residential heating appliance is maintained to a level where it can not only achieve the intent of its original design but also potentially exceed these expectations in performance and longevity.

The purpose of this book is not to replace existing manufacturer's maintenance directives but rather, to make the technician aware of their importance and the need to be knowledgeable of the appliance's ongoing maintenance needs throughout its service life.

John Certuse, P.E.
May 2019

ACKNOWLEDGMENTS

The National Academy of Forensic Engineers
The Old Colony Trade School
Lynn Manufacturing
Robinair Inc.
Testo Inc.
Fields Controls
Weil McLain
Bosch Inc.
Riello
R.W. Beckett
Carlin Combustion
Watts Industries
Honeywell © 2019 Resideo Technologies, Inc.
Rectorseal
Energy Kinetics
U.S. Boiler Company
McDonnell & Miller–Xylem Inc.
Taco Controls
Mastercraft Inc.
W.H. Riley Oil Company
Germain Plumbing & Heating
Robert Hunter
Judy Bass
Patricia Wallenburg

Judy Duguid

Lisa, Alec, Max

Photographs of manufacturers' items cited in credits.

Illustrations by John Certuse, P.E., unless otherwise stated.

Heating Systems
Troubleshooting & Repair

CHAPTER 1

MAINTENANCE THEORY

Everything, including any heating boiler or furnace, requires some level of preventative maintenance. This includes a simple visual examination of the appliance to ensure that there are no malfunctions (impending or present) and no interaction with debris or other undesirable condition that would jeopardize the appliance's safe and efficient operation. The saying "If it ain't broke, don't fix it" doesn't work here! Instead, it invites equipment and property damage as well as liability because of simple negligence.

As far as heating boilers and furnaces go, whether they burn oil, propane, or natural gas or use electricity to generate heat, recommended maintenance is usually (with few exceptions) annually, at the start of the heating season.

Are You Really Maintaining Your Equipment or Just Fixing It?

What does it mean when a machine like a boiler or furnace has been "properly maintained?"

Does this mean that it has been repaired when it broke down to make it operational again, or does it mean that it has been maintained to ensure a greater level of reliability without any unexpected breakdowns?

Does "properly maintaining" equipment mean more than just fixing it when it breaks, or does it also translate into maintaining optimum equipment performance and extend, perhaps indefinitely, the appliance's useful life?

The Consequences of "Run It Until It Doesn't Work Anymore"

Heating equipment that is inadequately maintained is likely to have a more severe breakdown, preventing it from operating or resulting in the need to be replaced. Also, you can bet that this equipment breakdown will occur on its schedule and *not* yours.

Imagine hosting a party at your home with company expected to arrive in half an hour, when you notice that the house is cold, and then your spouse asks, "Did you know there's water in the basement?"

Furthermore, if the heat needs to be restored quickly, you may not be able to get a serviceperson as quickly as you would like or need.

Delaying a routine and inexpensive cleaning of the air-conditioning condensate pan from the previous year can result in water damage and corrosion holes in your furnace's heat exchanger, making it necessary to replace your furnace, ductwork, AC evaporator, and condenser. This can cost as much as $12,000!

And if a service technician fails to perform maintenance actions that are of a "proper" standard practice of the trade, are common sense, or are recommended by the manufacturer, the company may find themselves liable for causing the damage to the appliance, as well as for causing any other damages extending from its failure.

Relying on a "run-to-failure," or *reactive maintenance*, approach invites inconvenience, increases costs, and risks potential liability as well as building damage (and more) when dealing with combustion equipment. To maintain a consistency in operation, reduce costs, eliminate building damage, and lessen the possibility of bodily injury to a house's occupants, predictability of impending failure and needed repairs must be the goal of not only the service company, but also the occupants of the houses in which the appliances are installed.

Building owners and homeowners must help service companies by allowing access to the equipment and scheduling repairs as well as giving the companies approval to perform required maintenance actions to keep the appliances operating in an effective and safe manner. "Kicking the can" down the road by delaying needed maintenance actions will only increase appliance damage and the likelihood of a more disruptive event in the future.

Technicians and service companies as well must become proficient in not just new equipment installation but also repair techniques of existing equipment. Leaving an older boiler or furnace to operate while in need of a repair can, in time, result in a more severe problem for both the homeowner and service company.

Installation to Code and Ongoing Maintenance

With residential heating appliances, installation and ongoing preventative maintenance are typically directed by the manufacturer's maintenance and installation manual, as well as applicable local building, mechanical, and plumbing codes.

With rare exceptions, the mechanical, gas, fuel, fire prevention, or oil burner code in effect for the state, province, or town where the appliance is located directs the *installation conditions* that must be followed for this equipment as well as auxiliary components such as fuel tanks and piping, combustion gas discharge, and fire safety systems.

Occasionally, codes mention ongoing maintenance. However, codes that originate from national organizations such as the International Code Council (ICC) and National Fire Protection Association (NFPA) primarily focus on the "installation conditions" and defer to the manufacturer's instructions regarding ongoing maintenance after the appliance has been installed.

Building codes, including those written by national organizations such as the NFPA, are only model codes or guides and are not enforceable until the local governing authority adopts them by statute. At that point the model code with amendments becomes law.

The governing authority may amend the code to make it more restrictive than the model code but not lessen it in it's influence. Changes may be made because of local conditions or events that have motivated local building and mechanical inspectors to implement additional restrictions in the installation of heating appliances. For example, safeguards may be mandated for a specific heating system, such as the installation of a fire protection sprinkler head as seen in Figure 1-1.

FIGURE 1-1 Fire Sprinkler Head Added to Boiler Installation

Three Levels of Maintenance

The three levels of maintenance used to keep a machine in proper working order are reactive, preventative, and predictive maintenance actions.

- Reactive maintenance—reacting to an unexpected equipment breakdown. If the equipment breaks down or displays an unacceptable failure such as a noise, emission, or other condition, the consequences of the failure are unacceptable and cannot be ignored, and they will require an immediate response.

Furthermore, the frequency of needed immediate or emergency repairs through reactive maintenance increases the more that preventative maintenance is avoided.

Example:
Suppose, due to a lack of routine cleaning, soot obstructs the flue passageways of a heating appliance, triggering the safety controls to shut the appliance off and thus requiring an immediate service call and cleaning—*a reactive maintenance action*. In addition, the soot blockage can also cause the appliance to overheat and damage various components. More alarmingly, using this example, if safety controls do not manage to shut the appliance down, poisonous carbon monoxide can be generated, which would then flow into the home.

- **Preventative maintenance—done before a more catastrophic failure has occurred.** Preventative maintenance actions are done with more predictability, typically during scheduled maintenance actions. Preventative maintenance actions include the replacement of consumable components, cleanings, and the visual assessment of equipment conditions.

  Sometimes a visual assessment of vital components and perhaps a simple cleaning of the components are all the maintenance that is required, and performing actual maintenance or parts replacement may not be necessary at all.

  Also note that the preventative maintenance actions of replacing consumable parts, doing routine cleanings, and visually assessing the vulnerable areas of the heating appliance should outnumber reactive maintenance emergency repairs and installations of new equipment. The goal is to maintain the boilers and furnaces so they may operate to their full life expectancy at a sustained maximum efficiency. As such, all aspects of an appliance's design, construction, and operation should be understood so as to be able to execute an effective preventative maintenance plan.

  Examples:
  Visually examining gas burners to check for corrosion, overheating damage, or dust and debris accumulation can prevent a burner malfunction that could lead to a more serious failure at a later date.

  Changing the filter and oil burner nozzle and pump screen on fuel oil systems will prevent fuel oil impurities from entering the oil burner and disrupting combustion that would cause soot production and distorted oil burner flames.

  Cleaning boiler and furnace flue passageways will enable future reliable and enhanced performance of the combustion appliance.

- **Predictive maintenance—assessing what maintenance will be required in the future.** This level of maintenance typically makes use of reactive and preventative maintenance actions and observations to predict what maintenance actions will be needed, and when, in the future life cycle of the equipment.

Example:
Suppose excessive flue passageway (or fireside) soot accumulation is noticed during annual preventative maintenance servicing or during an unexpected response to a no-heat call due to the appliance shutting down. The frequency of needed cleanings from the previous annual 12-month service event can be increased to include cleanings every 8 months to avoid unscheduled shutdowns and equipment failure.

The goal of utilizing the three levels of maintenance is to choose the maintenance action that lessens the impact on the homeowner in areas of cost and equipment downtime. Of the three maintenance action levels, it is best to reduce the frequency and impact of the costlier unplanned reactive maintenance actions during the service life of the appliance and to strive toward a maintenance approach favoring the preventative and predictive maintenance elements.

The impact and cost of performing a moderately involved maintenance action such as replacing the coil gasket of a boiler's domestic water heat exchanger in January is significantly more than performing the same job in July. The ability to *predict* the time before unrepairable damage occurs and schedule the repair when it is most cost effective and convenient is just good maintenance practice for the homeowner and supporting maintenance technician.

The Maintenance Mix

What proportion of maintenance should be reactive, what proportion should be preventative, and what proportion should be predictive? The ideal maintenance mix may vary among different types and ages of appliances. The maintenance mix is intended to reduce the risk of equipment failure, reduce operating costs, maximize equipment longevity, and make more efficient any ongoing maintenance in the future.

Using past experiences with actual as well as similar equipment components, you can build a maintenance plan that addresses impending failure in the most cost-effective and reliability-focused manner. Your maintenance plan should include two kinds of action steps: time-based and conditioned-based:

- **Time-based maintenance** consists of maintenance actions done at fixed intervals of operating time. An example of time-based maintenance is maintenance that is scheduled to be done during annual service inspections. This usually includes actions that are lower cost or that are a known necessity such as replacing fuel burner nozzles and filters and vacuuming and brushing the flue passageway.
- **Condition-based maintenance** is a repair action that depends on equipment assessments done at variable intervals. Condition-based maintenance relies on the visual assessment of the equipment and its components during regularly scheduled preventative maintenance. Examples of condition-based maintenance include gasket replacements that display a breakdown of material quality and the beginning stages of leakage.

P-F Curves and P-F Intervals—Predicting the Time to Failure

Many equipment components exhibit signs of impending failure prior to an actual catastrophic failure. This gives maintenance technicians the opportunity to make proactive repairs prior to equipment failure or damage

Maintenance engineers utilize what is known as a P-F curve (see Figure 1-2) to show the behavior of equipment, beginning with the initiation of component failure (point P on the graph), which precedes the ultimate mechanical failure of the equipment, that being point F on the graph. The time interval between these points is obtained by past experiences with actual as well as similar boiler or furnace components and is known as the P-F interval. This gives a technician an indication of the allowable time that a reactive maintenance action can be delayed for repairs.

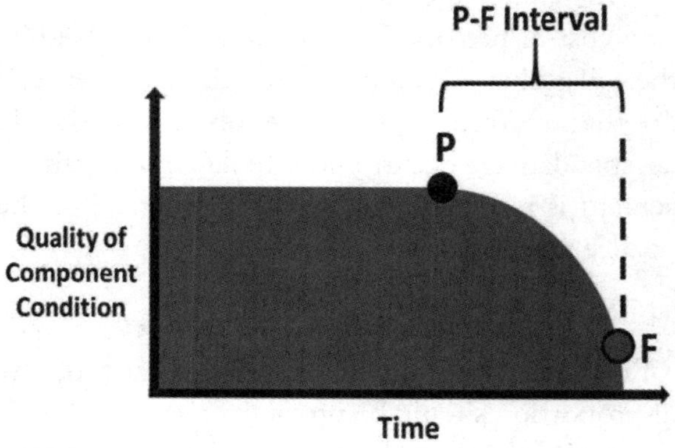

FIGURE 1-2 P-F Curve

Furnace and Boiler Life Expectancy

There are many factors that contribute to the life expectancy of a boiler or furnace. These include the initial design of the appliance, the conditions of installation, and the appliance's service history.

According to recent home inspector publications, the life expectancy for a furnace is 15 years, and for a boiler, it's 20 years. Yet there are appliances older than this that are still in good operational condition.

In fact, there are boilers and furnaces in use today that are over 100 years old. No, they are not as efficient as modern boilers and furnaces, but they are operational and perform their intended function in heating the building in which they are installed. Many of these older heating appliances, like the one shown in Figure 1-3, were designed and built in a way that they could function indefinitely if they were maintained, and no catastrophic event such as overheating or other material damage occurred. Much of this has to do with the materials that were used and the manner in which the appliances were built, making them more resistant to operational stresses and age-related issues.

FIGURE 1-3 Gravity Hot Air Furnace (courtesy of "The Old House Guy")

Why Replace a Boiler or Furnace?

What prompts the decision to replace a boiler or furnace? Think about these questions:

- Who advises that it be done, and what are the motivations for this advice?
- Are boilers and furnaces simply replaced because they reach some magic number that renders them incapable of operating, or is the decision based upon some assessment of their actual material condition or other factors affecting their operation?

In making a decision of whether to repair or replace, now consider the following.

Logistic Supportability

Traditional heating appliances that have been in use over the years have several sources to which a technician can go to obtain spare parts. Additionally, many of these spare components could be used in many different heating appliances regardless of the system, appliance type, make, or model.

The ready availability of parts for controls and consumable items such as refractories, seals, and gaskets is an expected level of support from the manufacturer or aftermarket parts suppliers. The level of support for spare appliance parts, however, tends to diminish when the needed parts move from the control and consumables realm to major components of the appliance's construction.

Spare parts for boilers and furnaces of a limited manufacturing run, or of a unique design, may be more difficult and time consuming to obtain. If a service company has the technical ability to fix an appliance but the required spare parts are several days or even weeks away, this may prompt the decision to replace the entire appliance as opposed to repairing it, especially in times of need such as during harsh weather conditions.

Difficulty of Repair

Anything "can" be fixed, but some technicians may not want to make "difficult" repairs due to the fear of liability, a lack of confidence in being able to perform the repair successfully, or a fear of continued unforeseen repairs not yet identified. And along with these fears and an unfamiliarity of the task at hand comes increased time in doing the repair.

Not Cost Effective to Repair

For smaller, less expensive boilers and furnaces, the cost of removing an appliance from system piping or ductwork and/or disassembling it to replace a furnace's heat exchanger or a boiler's section gasket, may approach the cost of purchasing a new appliance even if the manufacturer's warranty alleviates some materials costs. Also, considerations such as the cost of parts and labor and equipment downtime, as well as an estimate of successfulness, should be taken into account before the repair begins.

Increased Appliance Efficiency and the Return on Investment

Some people will replace an operating boiler or furnace because they believe that an increased operating efficiency will pay for the upgrade in a short time. Typical residential energy efficiency modernization programs include the addition of insulation as well as the occasional replacement of the heating boiler or furnace by a more efficient modern appliance.

Aside from adding insulation to a home to increase energy efficiency, how much money can be saved by upgrading to a more efficient boiler or furnace, and how long would it take to recoup that investment?

For a home equipped with an 80% efficient AFUE (annual fuel utilization efficiency) furnace during a hypothetical heating season that contains 5,949 heating degree days (HDD), 984 therms of natural gas will be used to maintain the home at a constant temperature of 68°F. This results in a *K factor* of 5,949 HDD/984 therms, or 6.04. The reciprocal of the K factor, known as the *burn rate*, is 0.165 therm per HDD. At a cost of $1.20 per therm of natural gas, the heating fuel costs for this particular winter are $1,180.

If the 80% efficient appliance were replaced with a series of furnaces of increasing AFUE efficiencies of 92, 96, 97, or 98% (for comparison purposes), under the same weather and indoor temperature conditions, how much of a cost savings would be realized by these different furnaces? Also, if fuel costs remain relatively level, can any projections be made about the need to replace the appliance due to efficiency differences alone? Table 1-1 presents the comparison.

TABLE 1-1 Cost-Saving Comparison of Upgrading from 80% Efficient Furnace for Heating Season (Winter of 5,949 HDD and fuel cost of $1.20 per therm. Cost of replacement furnace $5,500)

Furnace AFUE Efficiency	K Factor, HDD/ Therms	Burn Rate, Therms/ HDD	Amount of Fuel Used per Heating Season, Therms	Fuel Costs per Heating Season	Difference of Fuel Use Against 80% Furnace per Heating Season, Therms	Cost Savings Compared with 80% Furnace per Heating Season	Assuming $5,500 Furnace Cost and Fuel Cost Same, Years to Pay for Investment
80%	6.04	0.165	981	$1,180	—	—	—
92%	6.88	0.145	862	$1,035	119	$142.80	38 years
96%	7.2	0.138	820	$985	161	$193.20	28 years
97%	7.3	0.137	814	$977	167	$200.40	27 years
98%	7.39	0.135	804	$965	177	$212.40	25 years

At the reported expected lives of 15 and 20 years of heating equipment, even if the fuel costs doubled from those listed in the table, a new, more efficient appliance will not pay for itself. Also, regardless of these expressed life expectancies, proper installation and maintenance of combustion appliances can certainly extend their useful life, perhaps indefinitely.

Cost of Repair Versus Cost of New

Say a residential 10-year-old boiler has been shutting down due to products of combustion rolling out of the combustion chamber, triggering the rollout safety switches to trip and shut down the burners. The problem occurred whenever pieces of the deteriorating refractory that lined the firebox fell onto the burners, disrupting the combustion flow path. This also caused overheating of two of the five gas burner tubes.

The combustion chamber refractory is part of the manufacturer's construction components and will require removal, disassembly, component replacement, and reassembly of the boiler. The burner tubes and the base assembly that contains the refractory panels are available from the manufacturer in a reasonable time.

Provided that the repair is scheduled at an acceptable time of the year, the cost estimate to make the repair is as follows.

Repair Approach

Parts

- Boiler base assembly (refractory panels included): $342.83
- Main burner: 2 needed @ $49.50 each
- Total parts: $441.83

Labor at $110 per Hour

- Remove boiler controls, jacket, and necessary piping: 2 hours

- Remove failed base assembly: 2 hours (2 people, 1 hour each)
- Replace base assembly and burners and reassemble boiler: 3 hours
- Total labor: 7 hours at $110 per hour = $770 labor

Total Cost to Repair

- Labor and materials: $1211.83
- Profit and overhead, 20%: $242.36
- Total cost: $1454.20

Replacement Approach

New Boiler

- Cost of new boiler: $2,699.95

Labor @$110 per Hour

- Remove boiler from piping, replace boiler manifold and near piping, and wire boiler controls: 24 hours
- Total labor: 24 hours @ $110 per hour = $2,640.00

Total Cost to Replace

- Labor and materials: $5,339.95
- Profit and overhead, 20%: $1,067.99
- Total cost: $6,407.94

Whether to spend $2,642.20 to repair the boiler compared with spending $6,407.94 to buy and install a new one is a decision that must take into consideration the time, motivation, skill level needed to make the repairs, parts availability, and other factors. However, it should be realized that even major components of boilers and furnaces are available from the manufacturer and are supplied with the intent that their replacement is an expected field repair by technicians servicing the manufacturer's equipment. Furthermore, many manufacturers provide warranties for many of their appliance components, reducing the costs of these repairs.

Larger Commercial Versus Smaller Residential Boiler Repair

The size differences between a smaller residential boiler and a larger commercial boiler will affect the decision to repair or replace a failing appliance, since with the increase in size, there is an increase in costs and labor hours.

As an example, suppose that the boiler sections in a sectional cast-iron boiler are leaking.

Larger commercial boilers are delivered in sections. It is up to the installer to deliver these sections and assemble them at the site, as well as insulate and install piping and controls to make the boiler operational. Each section of these larger appliances may weigh several hundred pounds, and the boiler may contain up to 12 or more of these sections (see Figure 1-4).

In the event of a leaking fitting, gasket, or section, the boiler must be disassembled; the point of leakage identified, replaced, or repaired; and these large boiler sec-

FIGURE 1-4 Commercial Cast Iron Boiler Sections to be Delivered and Assembled On Site

tions reassembled to make the unit operational again. This requires multiple personnel to rig and assemble these components.

Now suppose a typical residential boiler has the same leaking condition as the commercial one described above. However, a replacement boiler is small enough to be delivered to the property completely assembled, allowing a single or smaller number of installers to make only minor piping and wiring connections, diminishing the repair option for these smaller units.

The Degradative Effects of Ignored Preventative Maintenance

Most situations where an appliance has failed and requires replacement could have been avoided had the appliance been serviced in a timely and effective manner.

Ignored Water Leaks

Ignored water or condensate leaks from boilers or furnaces can start out as inconsequential cosmetic damage. However, as time goes on, the rate of leakage as well as the corrosive damage to these appliances extends to other components and controls. This will result in additional damages, further increasing the costs of repair.

Examples:

- As water leaks from an appliance, evaporation increases the concentration of dissolved minerals in the water, which in turn can increase the corrosiveness of the leaking fluid.
- Water damage to controls and electrical connections can create short circuits and heat generation connections, causing unwanted incidences such as control malfunctions, equipment damage, unexpected shutdowns, and fires.

- Water leaks from air-conditioning coils installed in furnace plenum ductwork can lead to corrosion damage to furnace heat exchangers, inviting a combustion gas as well as a carbon monoxide leak into the system air.

Damage Due to Leaking Combustion By-Products

In addition to the effects of leaking water damaging a boiler or furnace, leaking combustion gases should also be identified and the leak fixed before equipment damage and contamination of the indoor air quality of a home occur.

Soot and other products of combustion, including carbon monoxide, that are introduced into the indoor air supply, as well as leaking combustion gas paths, decrease the performance of a boiler or furnace and cause damage due to the accumulation of combustion gas substances. Combustion gases can be corrosive, and for higher-temperature products of combustion, heat-induced corrosion or oxides can be generated, causing metal depletion and component wear.

Unintended gas path leakage also heats components of a boiler or furnace that were not designed nor intended to receive heat to this degree, requiring component repair or replacement. See Figure 1-5.

FIGURE 1-5 Clogged Boiler Allowing Flame Spillage from Observation Door

Leaking combustion gases can easily be detected with gas monitoring detection meters that react to the gases that are present in combustion by-products. While the system is in operation, downstream air in the furnace ductwork can be tested to identify if products of combustion are escaping from the furnace's internal heat exchanger. Likewise, the air space between a boiler's jacket and heat exchanger can be monitored to identify any combustion gases leaking from the boiler rope gaskets (Figure 1-6).

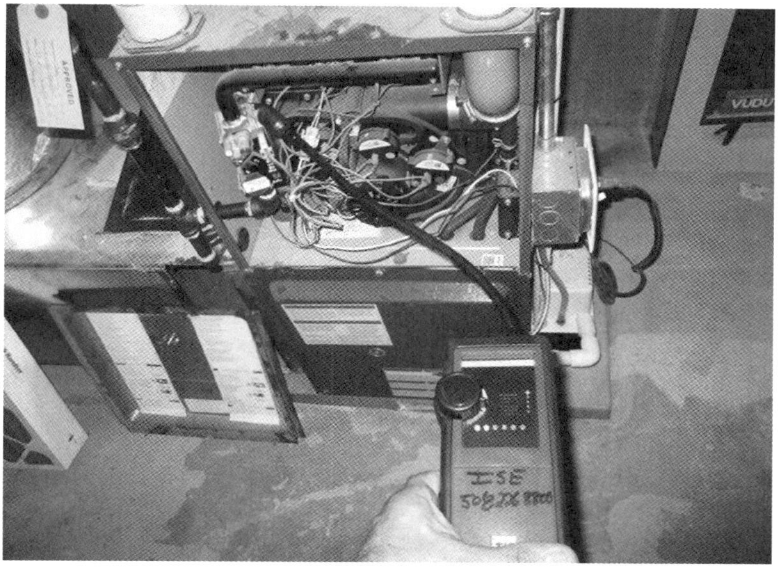

FIGURE 1-6 Detecting Combustion Gas Spillage (courtesy of Bosch)

The Building or Homeowner, the Technician, and the Manufacturer Working Together!

Once an appliance is installed and operational, what are the goals of the homeowner, the service company, and the manufacturer? All would agree that optimum equipment performance, reliability, and equipment longevity are in the best interest of everyone.

The deviation from this goal, however, usually begins when there is a breakdown in the commitment to maintain these appliances by any party, resulting in halfhearted efforts and compromise.

When a building owner asks, "Can it get by for this heating season?" when faced with an unexpected recommendation for a major repair, pressure is put on the technician and the service company to risk their reputation as well as face liability if building or appliance damage results due to this neglected maintenance.

The "cradle-to-grave" life-span of the equipment can sometimes be extended indefinitely provided that the appliance is not damaged beyond repair. Haphazard preventative maintenance can and likely will cause premature failure, resulting in unreliable operation, increased repair costs, and potential building damage.

Service technicians will spend most of their careers maintaining older systems as opposed to installing new ones. This being the case, they should and must be versed in the design of the the appliances as well as being familiar with preventative maintenance and inspection procedures needed to perform the maintenance to prevent equipment breakdown between maintenance intervals.

Likewise, manufacturers must provide educational literature and forums as well as parts to support their equipment's needs for consumable components such as filters, gaskets, and refractories along with those components that have an identified higher failure rate.

Finally, it falls on the homeowners to diligently schedule and allow technicians to perform needed maintenance on their systems. Haggling about price and hand-tying service professionals who only want to provide the homeowners with a reliable and safe appliance not only puts the boiler or furnace into a more precarious condition, but it also exposes service providers to liability in the event of a foreseen and avoidable equipment malfunction.

Reference

Moubray, John, *Reliability-Centered Maintenance*, Industrial Press, New York, 1997.

CHAPTER 2

HEATING SYSTEM TYPES AND PREVENTATIVE MAINTENANCE

A heating system generates heat from a boiler or furnace and distributes it throughout the home in measured, consistent proportions, maintaining consistent and comfortable temperatures.

The story of the evolution of the modern heating system if written would be enormous. Briefly, let's just say that it began as an open fire around which people would gather for warmth and for cooking. This open-fire approach evolved into the chimney-fireplace system.

There are examples through ancient times of milestones in the development of more advanced heating systems. In about 1300 BC, in Beycesutan, Turkey, King Arzawa directed that an underfloor fire-heated, radiant panel system be installed.

This design was later adopted by Roman builders when they constructed hollow floors with an internal flue passageway known as a hypocaust, through which wood-burning combustion gases would flow, heating the stone floor above (Figure 2-1).

Later, a combination of the fireplace–hot air system was designed by France's Louis Savot, who invented a circulating fireplace in the early 1600s. This innovation consisted of a fireplace with a hollow iron bottom and back that allowed natural convection air circulation and heat distribution.

Over time and likely in response to fires and the quest for convenience, by the end of the nineteenth century, a centrally located furnace or boiler was frequently installed to heat water or air or to make steam. The steam was distributed through the piping or ductwork of the building's heating system to provide heat throughout the building.

FIGURE 2-1 Roman Hypocaust (Courtesy of GNU Free Documentation)

Older Systems Still in Use Today

Throughout the world, older buildings, and hence older heating systems, still exist and are still being used today. The heat distribution systems of piping or ductwork that carry steam, hot water, or hot air are part of the buildings' original construction and infrastructure. The original system may be connected to the same boiler or furnace that was installed when the house was built, or it may be connected to a newer replacement heating appliance. Regardless, the heat distribution system dictates what type of boiler or furnace and control configuration is integrated into it as well as the type of maintenance the appliance and heating system require.

Remarkably, some of the systems that were designed and installed over a hundred years ago are still in use today. They are described briefly below.

Gravity Hot Air

Gravity hot air heating systems were installed in the late 1800s and into the 1940s and utilize a furnace as well as supply and return ductwork.

An example of the simplest form of a gravity hot air heating system is the wall or floor furnace. This heating appliance incorporates a heat exchanger in which heat from combustion gases is transferred to system air, which is distributed by natural convection circulation methods.

Wall and floor furnaces do not distribute heat through any ductwork system but rather directly discharge heated air into the room where they are installed (Figure 2-2).

Natural convection transfer of heated air throughout a home is also accomplished on a greater scale by the use of a central heating furnace, known as the gravity hot air heating system. Heated air is distributed by natural convection, where warmer air heated through the central heat exchanger rises and passes through the supply ductwork to various rooms within the home.

Likewise, cooler and denser air is returned to the furnace through floor registers connected to return ductwork, completing the closed loop of this system (Figure 2-3).

FIGURE 2-2 Floor Furnace

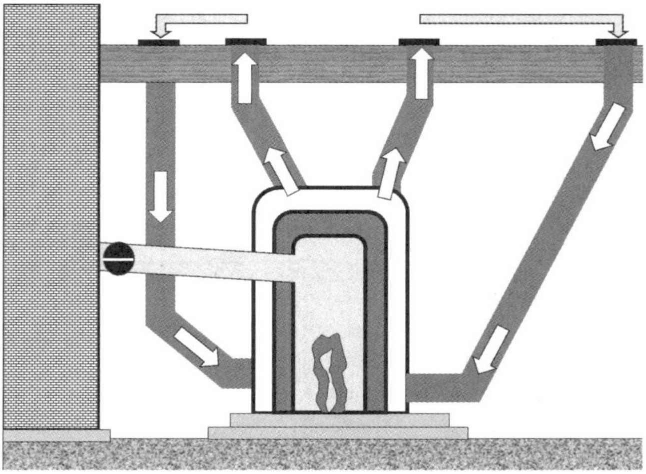

FIGURE 2-3 Gravity Hot Air System

Unlike forced hot air furnace systems, gravity hot air furnaces do not use a blower fan to create airflow. Instead, they contain many individual ducts connected directly to the furnace, as opposed to a ductwork system consisting of a main trunk and individual branches. This gives the gravity hot air furnace the appearance and nickname of the "octopus furnace" (Figure 2-4).

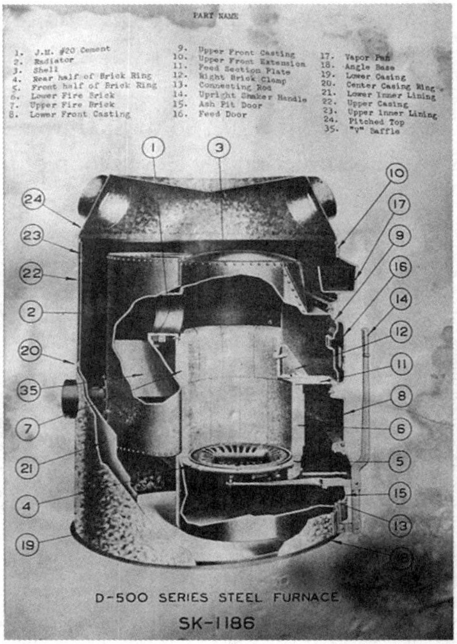

FIGURE 2-4 Gravity Hot Air Furnace (Courtesy of "The Old House Guy")

Gravity Hot Water

A gravity hot water boiler and system is an early type of hydronic heating system that does not rely upon a circulating pump for water flow. It accomplishes heat transfer throughout the system's cast-iron radiators by natural convection induced circulation (Figure 2-5).

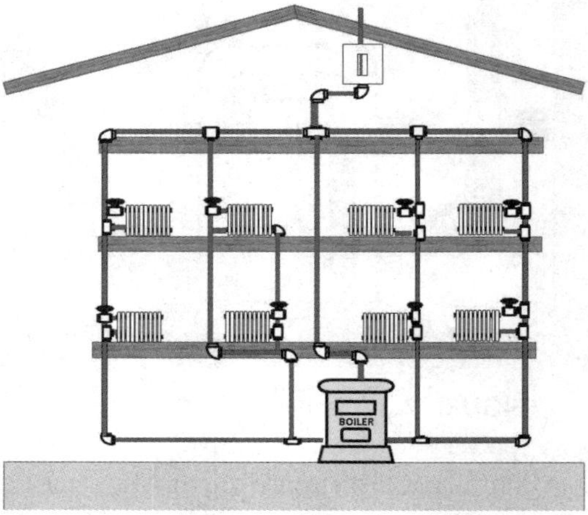

FIGURE 2-5 Gravity Hot Water System (Courtesy of heatinghelp.com)

This type of system consists of a central boiler and supply and return piping that is larger in diameter than modern *forced* hot water system piping. The larger diameter of the piping, which is typically in the 3- to 4-inch range, allows a greater volume of water to flow through the system, which is necessary due to the lower natural convection flow rates.

Heat distribution piping in gravity hot water heating systems makes up a circuit of system water flow facilitating the unimpeded flow of hot water to and then from cast-iron radiators in the various rooms of the house. To maintain water under pressure, a head tank is installed at the upper region of the system, usually located in the home's attic.

Initially, these systems did not have low-water safety controls. Due to the large volume of water in the boiler and large-diameter piping, low-water issues were noticed when the system no longer heated adequately, indicating that an insufficient volume of water was present in the system. To avoid the low-water problem, the boilers in many of these older gravity hot water heating systems have been retrofitted with more modern auxiliary components such as automatic water feeder valves.

Steam

The steam heating system consists of a centrally installed boiler that may have initially been a coal-fired appliance. The boiler is partially filled with water, allowing the upper region to contain a quantity of steam and the lower region to contain water, preventing an overheating event. From this a system of supply and return piping provides cast-iron radiators throughout the home with saturated steam, typically less than ½ psi in residential installations.

As the two-pipe steam system in Figure 2-6 shows, steam enters the various cast-iron radiators. The steam delivers its latent heat of vaporization within each radiator, where the steam then turns back to its water form, known as condensate. From here the condensate flows back to the boiler, completing and likewise beginning the cycle once again.

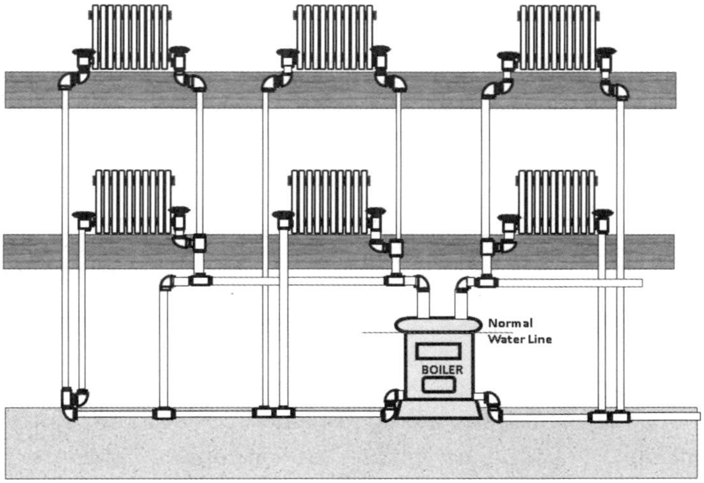

FIGURE 2-6 Steam Heating System (Courtesy of heatinghelp.com)

In the one-pipe system (Figure 2-7), steam enters into and condensate returns out of the same pipe connected to the radiator.

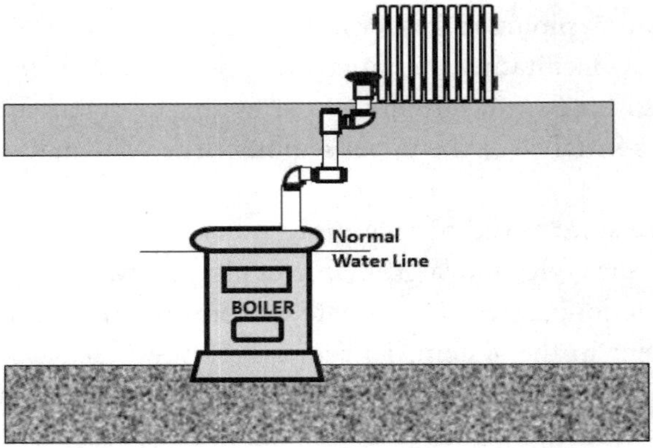

FIGURE 2-7 One-Pipe Steam System

Steam boilers are protected from low water or dry firing by the installation of low-water cutoff switches. Additionally, overpressurization is prevented by a pressure-regulating control as well as a safety relief valve.

Steam heating systems also employ air vents located on radiators and steam headers to allow air to escape from the system while retaining the steam generated in the boiler. Cast-iron radiators are also outfitted with steam traps that allow the radiators to retain steam while expelling cooler condensate, returning it to the system's return piping and boiler.

■ ■ ■

Gravity hot water systems, as well as gravity hot air and steam-heating systems, fell out of favor in the United States by the 1940s, and for the most part, the common heating systems in use today rely on forced hydronic water flow and convection.

These systems—for the most part found in residential structures—were initially built in a single-system configuration controlled by a centrally located, single thermostat. Outlying regions of the home did not always experience the comfort of those in the immediate vicinity of the main thermostat due to their distance from the thermostat.

MAINTENANCE TIP!

A Note About Early Heating Equipment and Asbestos

Many older boilers and furnaces were constructed with asbestos insulation, refractory components, and piping insulation. Learn to recognize asbestos in the event that it must be manipulated during repairs. Asbestos can be found in the form of loose-fill insulation, a dense chalk-like brittle substance, or a corrugated box-like insulating panel. See Figures 2-8, 2-9, and 2-10.

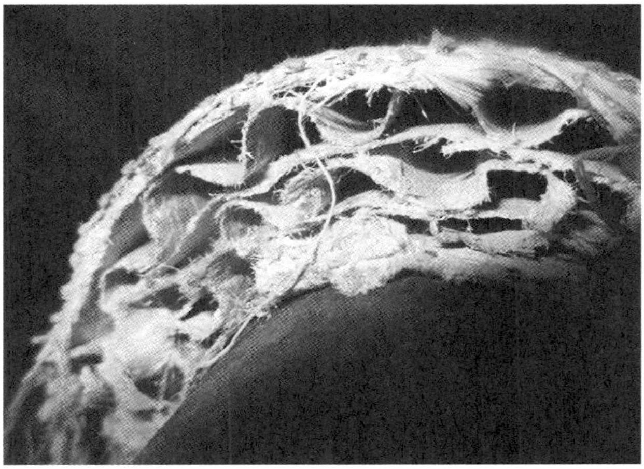

FIGURE 2-8 Corrugated Asbestos Piping Insulation

FIGURE 2-9 Loose Fill Asbestos Insulation

FIGURE 2-10 Corrugated Asbestos Panel Insulation

Never remove this material on your own, and always utilize a certified asbestos abatement contractor for this purpose! Careless actions by haphazardly removing this material from a boiler, furnace, or heating system piping may contaminate you as well as the home and expose unsuspecting families to these materials even years after the work is done.

Modern Heating Systems

Forced Hot Water (Hydronic)

The forced hot water heating system, also known as hydronic heating, utilizes hot water under pressure to act as the heat transfer medium between the boiler and heat exchangers throughout the home.

Forced hot water heating systems are composed of a boiler and a hydronic system of supply and return piping that is filled with water and contains no air or steam pockets. The system is designed to allow heated water to flow to baseboards (Figure 2-11) or radiators throughout the house on demand in response to calls for heat from the system's thermostat

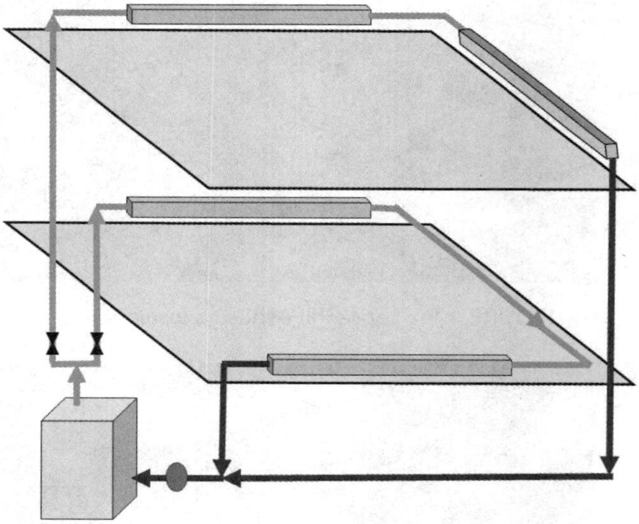

FIGURE 2-11 Two-Zone Forced Hot Water Baseboard System

Multiple "loops" of hydronic heating systems known as zones allow different temperatures to be maintained throughout the home's various areas, with each zone being controlled by an individual thermostat.

Forced hot water boilers utilize a control configuration of either a hydronic water circulating pump in conjunction with one or more thermostat-controlled zone valves or several thermostat-controlled circulating pumps regulating the flow of water through each zone.

Baseboards and radiators can also be replaced with forced hot water radiant heating systems, where water tubing is installed within floors (Figure 2-12) and in other building cavities. Radiant floor heating systems have the advantage of a greater floor layout without the need to position baseboards around the perimeter of the room. Also, a heated floor distributes heat more evenly throughout the room.

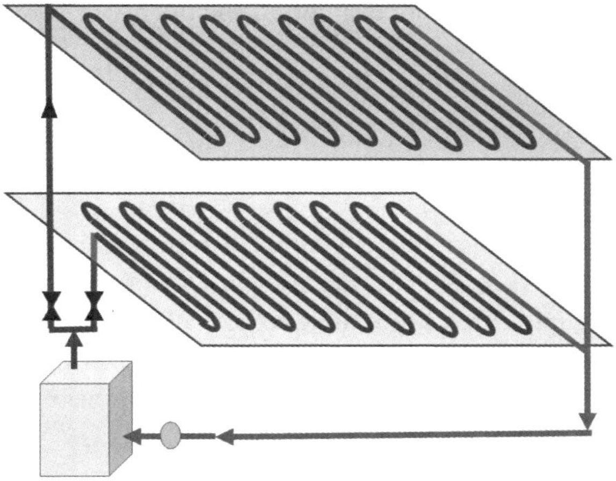

FIGURE 2-12 Radiant Floor Hydronic Heating System

Forced Hot Air

In a forced hot air heating system, a furnace distributes heated air through a system of supply and return ductwork. A circulating "blower" fan directs the air through the ductwork to and then from each room in the home (Figure 2-13). Zone separation of the ductwork system is accomplished by means of dampers that control the flow of air in branch circuits.

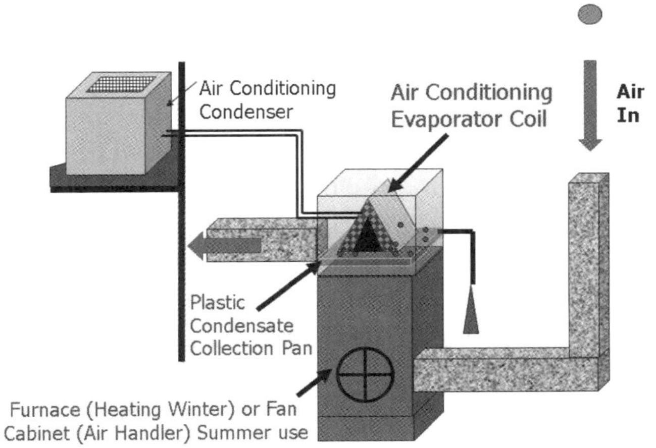

FIGURE 2-13 Forced Hot Air Furnace and System Ductwork Outfitted with Air Conditioning Components

Air is heated in the furnace by combustion gases passing within the flue passageways of the internal heat exchanger. This heat exchanger is located in the furnace cabinet. Heat is transferred when the furnace blower moves air through the furnace's cabinet and over the exterior of the heat exchanger.

Forced hot air heating systems are especially convenient because they act as air-conditioning systems as well. The dual operation is made possible with the installation of an air-conditioning evaporator coil in the ductwork plenum chamber at the furnace blower discharge.

Furnaces can utilize heating oil, propane, natural gas, or electric resistance heating coils as their energy source.

The Heat Pump System

In the heat pump system, warm air is generated from compressed refrigerant gases passing through a heat exchanger in the air handler blower cabinet during winter months. The heat pump utilizes a reversible refrigeration cycle to direct condensed or expanded refrigerant through evaporator or condenser coils to achieve heat generation or absorption, depending on whether the system is in heating or cooling mode.

Hydro Air

Not only can warm air be generated in a furnace or heat pump system. It can be generated from hydronic forced hot water boilers as well. The hydro air system utilizes a centrally located forced hot water boiler that supplies heated water to heat exchanger coils located either in the air handler blower cabinets or in the ductwork they are connected to. This type of system allows heat generation in the air handler without combustion occurring directly within this component of the system. This reduces the possibility of soot infiltration issues known as "puffbacks" and allows the flexibility of heat control of a forced hot air system.

This type of system is also attractive for larger homes since there is no need for multiple fuel-burning furnaces and their associated fuel and combustion exhaust systems. See Figure 2-14.

Also, air handler cabinets in hydro air systems can be positioned throughout the home, shortening the lengths of ductwork needed, eliminating air balancing problems, and allowing better heat control in each zone supplied by each unit.

The installation of evaporator coils in each air handler creates a dual heating and cooling operation of the system, eliminating the need for separate systems.

Preventative Maintenance Pertaining to the Heat Distribution System

Preventative maintenance is not limited to the boiler, furnace, or air handler alone. Oftentimes it is the *system* that distributes the heated steam, water, or air throughout the building that is underperforming due to needed maintenance. The entire heating system, including all piping and ductwork, should be considered and maintained and not just the system's individual heating appliance.

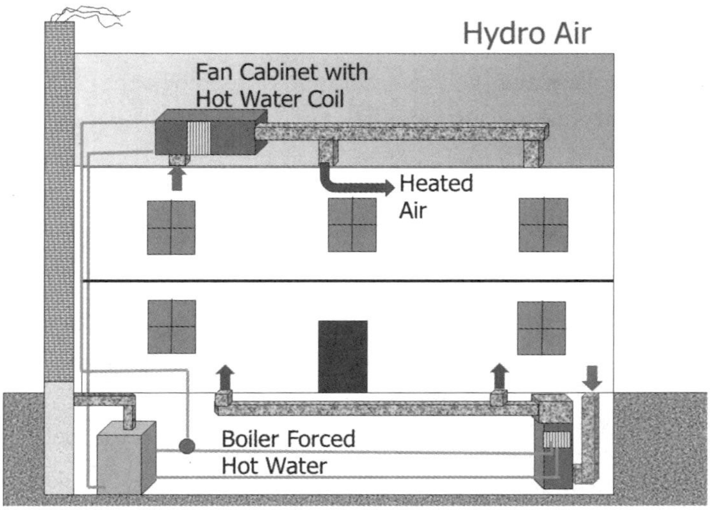

FIGURE 2-14 Two-Zone Hydro Air System

Aside from the preventative maintenance and inspection of the boiler, furnace, or air handler, heating systems require periodic maintenance as well. Areas of needed attention include water chemistry, cleanliness of the ductwork and heat exchanger fins, and repair of the auxiliary boiler, furnace, and system controls and components.

Some heating systems require more maintenance than others. It depends on the type of heating medium being produced as well as circumstances in the home such as the presence of pets or indoor air quality (IAQ) issues.

The local water supply, as well as the water in the boiler system, contains minerals of varying levels of hardness and other properties, which if left unchecked can have a detrimental effect on heating systems. These minerals vary depending on the water supply but likely will include silica, calcium, and magnesium salts. The higher concentration of these minerals the greater the ability of the appliance to generate sludge and scale as well as a corrosive environment to the boiler water.

Leaking boilers are common and often go unnoticed since the leaks may be in hidden locations. For example, boiler section connections might leak water into the combustion chamber, where the water will evaporate, and so there will be no accumulation of noticeable water.

Forced Hot Water System Maintenance

Heat transfer from a forced hot water baseboard to room air is accomplished by natural convection. In this cycle, cold air due to density differences flows into the bottom of the baseboard, where it is heated by the hot water in the copper pipe–aluminum-finned radiator. From here, the warm air rises, and by natural convection, it continues to recirculate to complete the cycle of heating room air to a uniform temperature until the thermostat is satisfied ceasing the flow of heated hydronic water through the system piping.

IAQ contaminant particles that are in the room air will also follow this path, causing them to pass through and, over time, cling to the aluminum fins of the baseboard (Figure 2-15). Eventually, these accumulations block the airflow through the baseboard element, decreasing system efficiency.

FIGURE 2-15 Dust Accumulation of Forced Hot Water Baseboard Fins

On a routine basis, baseboard and radiators should be cleaned by vacuuming any accumulating dust or debris that has collected. See Figure 2-16.

FIGURE 2-16 Vacuuming Baseboard Fins to Assure Unimpeded Air Flow and Efficiency

Since loss of or damage to the baseboard aluminum fins also reduces the thermal efficiency transfer of the baseboards, any impact damage should be addressed by repairing the bent fins. This can be done with a fin-straightening comb tool or a putty knife (Figure 2-17). Unrepairable baseboard sections can be replaced, and the damaged element can be replaced as well.

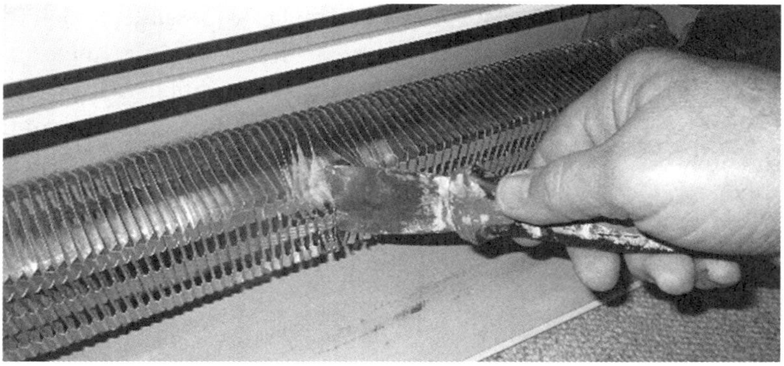

FIGURE 2-17 Straightening Baseboard Fins

Steam System Maintenance

Steam System Air Vents

Steam systems require significant preventative maintenance actions that if neglected can cause damage to the boiler supplying them with steam. These systems must be maintained to allow them to retain as much steam and water as they can to remain a "closed" system and not require excessive makeup water. Makeup water contains oxygen and other non-condensable gases that are detrimental to good boiler water chemistry and that must be purged from the heating system.

Another consequence of excessive loss of system steam and the addition of excessive makeup water to residential steam boilers is corrosion damage to the boiler. Excessive makeup water creates a corrosive environment in the boiler's upper steam holding area, resulting in corrosion damage to the upper casting sections of the boiler, as seen in Figure 2-18.

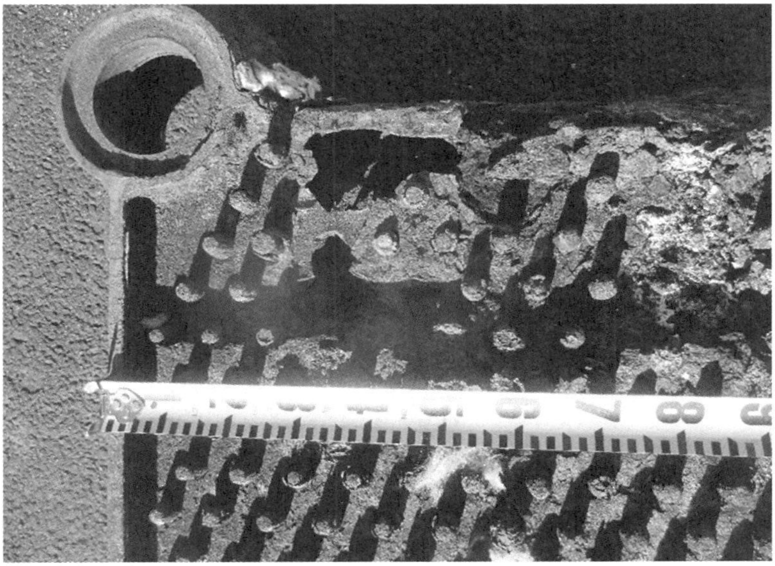

FIGURE 2-18 Oxidation Corrosion Damage to Boiler Section

So the goal is to minimize steam loss as much as possible, which means that steam vents should be inspected as a part of steam system maintenance.

Steam vents remove air from steam systems and are located on steam supply piping in the basement where the boiler is located as well as on radiators in the various rooms of the home.

FIGURE 2-19 Steam Vent on Cast Iron Radiator

Malfunctioning steam vents can be identified while in operation because they display excessive steam leakage (Figure 2-20). Malfunctioning steam vents can cause the abnormal loss of steam, which in turn requires the addition of makeup water to the boiler. The need to add a greater than usual amount of makeup water to the boiler may also be an indication that the system is losing steam from leaking vents. When a malfunction occurs, the vent will continuously leak air as well as steam, requiring more frequent makeup water addition.

FIGURE 2-20 Malfunctioning Steam Vent

Steam Traps

Steam traps are automatic modulating valves installed on steam system radiators to discharge condensate back to the boiler after its useful heat during the transition from steam has been spent (Figure 2-21). A steam trap may fail by letting all the steam pass through before its useful heat can be used in the radiator, or it may fail in a closed position and block any flow of condensate through the trap outlet.

FIGURE 2-21 Steam Trap Installed on Cast Iron Boiler

A leaking radiator's steam trap prevents all the steam from condensing in the radiator, allowing the steam to flow back to the boiler, thereby wasting its heat content and making the system less efficient. Furthermore, condensate mixed with steam in the system piping results in impact of the condensate particles and causes "water hammer" impact sounds.

Leaking steam traps can be identified using ultrasound amplifier detectors. A leaking steam trap presents a loud rushing sound downstream of the trap while in operation.

Infrared imaging is also used to determine if the trap is leaking or blocked based upon the temperature differences between the inlet and outlet.

Forced Hot Air Ductwork System Maintenance

Forced and gravity hot air systems heat by moving the air mass flow into and out of a furnace through a system of ductwork. This will also cause entrained household dust that consists of carpet, furniture, and clothing fibers along with mold spores, pet dander, and skin cells to be drawn in and remain inside this ductwork.

Furnace and air handler filters must be replaced as needed depending on the home's indoor air quality conditions, and over time, the ductwork must be cleaned (Figure 2-22). Failure to do so results in a decreased efficiency of airflow and can expose the home's occupants to increased levels of airborne particles.

FIGURE 2-22 Ductwork in Need of Cleaning

Air-Conditioning Condensate System Maintenance

For forced hot air furnaces, air-conditioning air handlers, or hydro air systems equipped with air-conditioning evaporator coils, a condensate disposal system is used to collect and drain away water generated by condensed humidity while the system is in operation during summer months. The condensate collection system consists of a primary condensate pan located under the refrigerant cooling coils. The condensate pan is connected to condensate drain piping pitched away from the pan, allowing the continual drainage of condensate water from the air-conditioning system (Figure 2-23).

FIGURE 2-23 Condensate Piping

Over time, dust, water hardness minerals, and microbiological growth will build up in the condensate pans and drain piping, slowing and eventually blocking the flow of condensate being generated by these systems when they are in operation. When this occurs, the internal condensate pans will overflow, which if gone unnoticed will cause water and mold damage to the areas in the building where they are installed. This condition will also cause the appliances to become wet, requiring cleaning, repair, or replacement (Figure 2-24).

FIGURE 2-24 Condensate (Water) Damage to Air Handler

The condensate system should be cleaned to remove algae, hardness minerals, and water-bearing microorganisms or debris from the evaporator condensate pan and drain piping. A wet vacuum or compressed air can be used to clear standing condensate from the system piping. Additionally, the condensate pan should be cleaned by opening the evaporator coil casing and clearing the pan of any accumulated slime or debris (Figure 2-25). Commercial cleaning chemicals and biocide solutions can be applied to stop microorganism growth as well as clean the evaporator coil and fins.

FIGURE 2-25 Fouled Air Conditioning Evaporator Coil, Condensate Pan

To keep condensate system traps in unheated attic locations from freezing and cracking, resulting in a condensate leak the following spring, at year's end add anti-freeze or clear the condensate lines with a wet vacuum or compressed air.

References

ACCA Manual J Residential Load Calculation.

Flynn, D., *The Nalco Water Handbook* (3rd ed.), McGraw-Hill, New York, 2009.

Hydronics Institute Guide H-22.

I=B=R Guide to Residential Hydronic Heating.

IECC.

IMC.

CHAPTER 3

INSTALLATION TO CODE AND ONGOING MAINTENANCE

Heating System Design

There are several ways to determine the sizing and design of the heating system. The calculation can be done by hand, utilizing heat transfer calculation methods and listed performance values of windows, doors, and insulation materials. There are also computer programs that calculate heating and cooling loads.

However, the first step is to check with local building officials about local requirements since there may be specific requirements about what evaluation method they will accept before issuing a mechanical permit approving the heating system intended to be installed. Requirements to improve energy conservation efficiency are being incorporated into many local building and mechanical codes, directing a cumulative, community-wide move toward residential structure modernization and increased energy-efficient homes. This directive begins with the design of more efficient heating equipment as well as properly designed and installed heat distribution systems.

Heat Load Determination

The design of a residential heating system must first address the home's requirements for heat and cooling output, which dictates the air-conditioning load as well as the net heat distribution from the boiler, furnace, or heat pump. The sizing of each room's required heat input under design conditions will indicate the hydronic forced hot water baseboard or radiant tube length as well as the forced air ductwork sizes needed. The engineering model specifying how the system is being designed to perform relies upon the design conditions identified such as those specified by the American Society of Heating, Refrigeration and Air-Conditioning Engineers (ASHRAE).

The approach for sizing heating systems is one where the building's dimensions and construction materials, such as framing thickness and insulation R values, are evaluated for the exterior abutting wall, ceiling, or floor, making up the heated (or cooled) "conditioned space." From this, the overall coefficient of heat transfer for that building is determined. The building's overall heat loss (or gain for cooling load calculations) is then applied to the design conditions of the area where the building is situated. This yields a final BTU/hr (British thermal unit per hour) requirement that a boiler or furnace must deliver to maintain the design's indoor temperature under the design's outdoor temperature conditions. Typically, the design conditions for heating are the expected coldest temperatures that will occur 99% of the time at 1% cooling dry bulb and 99% heating dry bulb conditions for the area where the building is situated.

In addition to methods for performing heating and cooling load calculations by hand and the automated computerized programs that do the calculations, there are also many heating and cooling system sizing guidelines such as the Hydronics Institute's *Guide H-22* and the Air Conditioning Contractors of America's (ACCA) *Manual J*.

Hydronic Heating System Design

Design guidelines for the layout and considerations of hydronic heating system piping are available in the *I=B=R Guide to Residential Hydronic Heating* published by the Hydronics Institute Division of the Gas Appliance Manufacturers Association (GAMA). This guide addresses issues such as boiler selection, placement, and accessories as well as heating system selection and piping system design.

Appliance Placement

Clearances to Combustibles

Combustion appliances are designed with an understanding of the heat they radiate and conduct while in operation. The possibility that they may cause adjacent combustible materials to catch fire under normal and malfunction conditions is considered in the design of the appliances and their recommended use.

Manufacturers' instructions will likely cite acceptable national and code requirements to be followed in the installation of their appliances, and the manufacturer and code guidelines may stipulate specific installation requirements for each appliance regarding clearances to combustibles. In addition, the National Fire Protection Association publishes national directives such as *NFPA 31: The Standard for the Installation of Oil Burning Equipment* and *NFPA 54: The National Fuel Gas Code.*

The potential for a fire caused by too small a required distance between a combustion appliance and walls or ceilings is typically not alleviated by construction modifications such as the addition of sheetrock or sheet metal or other noncombustible materials on top of wooden construction materials, unless the modifications are prescribed by the manufacturer or code guidelines. Combustion appliances can allow heat transfer through these outer noncombustible materials, which in turn transfer heat to the combustible materials beneath them. Furthermore, the applied materials themselves—for example, a concrete pad constructed over a wooden floor to attempt to compensate for proper clearance to the wooden floor—may fail over time and crack due to the proximity to the appliance's heat source, increasing the likelihood of a fire (see Figure 3-1).

FIGURE 3-1 Boiler Installed on Combustible Floor Contrary to Manufacturer's Instructions

And so while even a properly functioning appliance may operate for years or even decades in an installation where the directive regarding clearances to combustibles is violated, the potential still exists that a fire may develop in time. A normally operating appliance in close contact with a combustible material has the potential to degrade that material over time, making it more capable of being ignited even though no equipment malfunction has resulted.

In addition to the need to verify the installation clearances to combustibles of the appliance itself, the connector vent piping clearances to combustibles must also be considered.

Confined Space of Installation

Another important consideration in the installation of a combustion appliance is the quantity of air available to the appliance for combustion. Although many combustion appliances are designed to be installed with sealed combustion systems, i.e., piping combustion air passageways that deliver combustion air to the burner with corresponding combustion gas exhaust piping systems, many appliances utilize combustion air directly from the space in which they are installed.

The specific requirements of the equipment manufacturer and local applicable building and installation codes should be referenced. Generally, most manufacturers require 50 cubic feet of air per 1,000 BTU/hr of input for combustion appliances, no matter whether the fuel is heating oil, natural gas, or propane. Thus a 120,000-BTU/hr input furnace that is taking all combustion air from the room where it is installed requires 2,400 cubic feet of air to support combustion in accordance with the manufacturer's and code requirements.

Suppose, though, that the above appliance is enclosed in a room with less air, perhaps as part of a renovation project. For example, say the furnace is enclosed in a room that measures 20 feet by 15 feet by 7 feet in height (2,100 cubic feet). In this case, the appliance is in violation of these guidelines—the appliance is installed in a *confined space of installation* (Figure 3-2). Additional combustion air, however, can be supplied to appliances installed in spaces where the overall volume of air is too small to support their requirements.

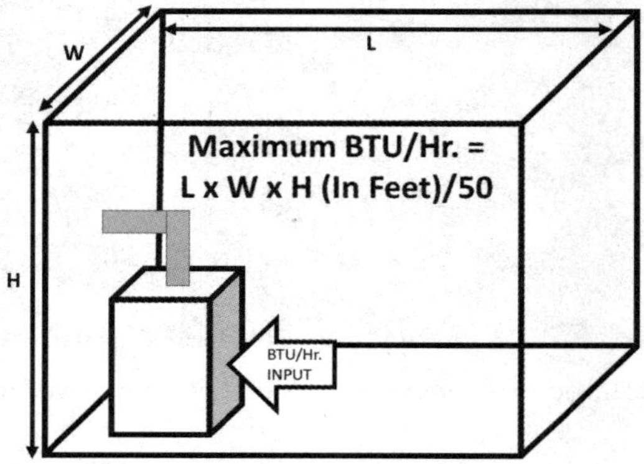

FIGURE 3-2 Appliance in a Confined Space of Installation

All Combustion Air from Inside the Home

If an oil-, natural gas-, or propane-fired boiler or furnace, as well as other combustion appliances such as a gas-fired water heater or clothes dryer, is installed in a confined space of installation, a means must be provided to supply that room with an adequate supply of combustion air.

Illustrations of acceptable methods are provided by local codes and specific equipment manufacturers. As an example, one such remedy is the installation of air openings to adjacent rooms to allow the air volume of those spaces to be made available to the combustion appliance.

When adjacent space room air is used to supply air to such an installation, two permanent openings must be made to the adjacent space or spaces so that the combined volume of all the spaces meets the appliance's air requirements. The openings are such that one opening must be within 12 inches of the top of the room and one within 12 inches of the bottom of the room. Each opening must have a free area equal to a minimum of 1 square inch per 1,000 BTU/hr of the input of *all* the appliances in the space *but* not less than 100 square inches.

All Combustion Air from Outside

Some combustion appliances are installed in a confined space that abuts an exterior wall, allowing the introduction of outdoor combustion air. These appliances have their own guidelines requiring access to this air to aid combustion.

Outside air utilized for combustion is supplied typically by ducts or openings, one opening within 12 inches of the top of the room and the second within 12 inches of the bottom of the room. These openings are permitted to be connected to any other spaces that are communicating to the outdoors, such as ventilated attics and crawlspaces. Any spaces that are connected to the two openings should not intersect, keeping the air passageways separate from each other, and the duct serving the upper opening should extend upward from the appliance space.

The air openings of vertical ducts must have a free area of 1 square inch per 4,000 BTU/hr of the total input of all appliances in the room. Each opening in horizontal ducts must have a free area of 1 square inch per 2,000 BTU/hr of the total of all appliances in the space.

Ducts should have the same cross-sectional area as the required free area of the openings to which they connect, and the minimal cross-sectional dimension of the rectangular air duct should be 3 inches.

Attic Combustion Air

Access to air in the attic space can also be used to provide the required volume of combustion air. The air is provided with a metal sleeve extending from the appliance enclosure to at least 6 inches above the top of the ceiling joists and ceiling insulation. An inlet air duct within an air duct is acceptable provided that the inlet duct extends at least 12 inches above the top of the outlet duct in the attic space. The end of the ducts that terminate in the attic should not be screened.

Under-Floor Combustion Air

Combustion air can also be obtained from areas under the floor provided that the openings have free opening areas to the outside equivalent to not less than twice the required combustion air opening listed for appliances obtaining air from the outside.

Mechanical Means to Provide Combustion Air

In the event that mechanical room openings or duct systems to provide combustion air are not possible or desired, combustion air can be provided by mechanical means (Figure 3-3) with the use of an air supply fan electrically interlocked with the burner of the appliance installed in the confined space.

FIGURE 3-3 Fields Fan-in-a Drum (Courtesy of Field Controls)

MAINTENANCE TIP!

Signs of Inadequate Combustion Air

For an appliance not receiving adequate combustion air, the lack of air will result in incomplete combustion. When there is incomplete combustion, the fuel is not being completely burned. This results in carbon accumulations (soot), which has detrimental effects on the operation of the appliance, including:

- Soot production
- Problems with the oil burner ignition transformer
- Interior moisture issues
- Interior odors
- Smoke soot damage
- Carbon monoxide
- Decreased draft
- Condensation

When there is an inadequate supply of combustion air, all fuel sources, not only oil but propane and natural gas as well, can produce soot. Soot causes obstruction of the flue and combustion gas pathways, causing combustion gas spillage outside the appliance, including water vapor, soot, and carbon monoxide.

For boilers or furnaces outfitted with oil burners, in the event of an obstructed combustion gas path, products of combustion will often discharge from the observation door or other firebox openings (Figure 3-4). Since the oil burner is typically installed in this area, this may cause heat damage to the ignition transformer and oil burner components.

FIGURE 3-4 Obstructed Combustion Gas Path Causing Products of Combustion to Flow Back from the Boiler's Observation Door

Chimneys and Direct Venting

Masonry

The traditional way that combustion appliances discharged products of combustion outside the structure was through a chimney. Traditional masonry chimneys are constructed with terra-cotta clay flue liners in a brick casing, with each flue allocated to one individual "appliance," whether it is a boiler, furnace, water heater, or fireplace (Figure 3-5).

Masonry chimneys can also be lined with metal flue liners. These liners allow ease of maintenance in the event of needed liner replacement over time.

Metal Flues

The construction of masonry, metal, and factory-built chimneys is directed in code guidelines such as NFPA 211's *Standard for Chimneys, Fireplaces, Vents, and Solid Fuel-Burning Appliances* as well as local codes and manufacturers' instructions.

Metal flues can be installed in masonry chimney casings as well as wooden enclosures as long as required diameters and specific clearances to combustible distances are maintained. Metal flues may also be installed as a stand-alone chimney, as seen in Figure 3-6.

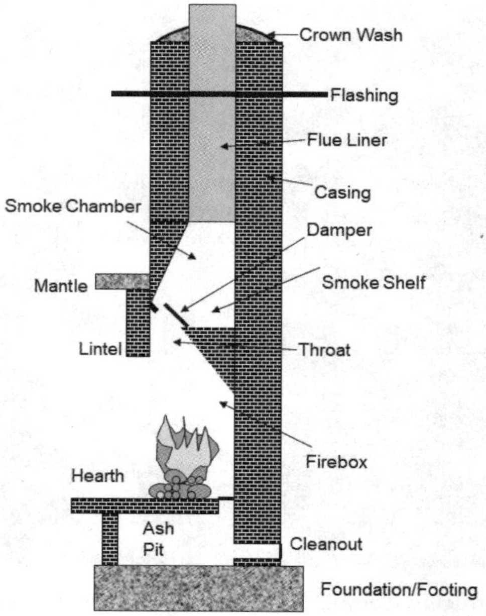

FIGURE 3-5 Masonry Chimney Construction Terms

FIGURE 3-6 Metal Chimney-Flue Added to House

Venting and Appliance Categories

Equipment is classified by the conditions in which the venting system operates. The venting system is defined as the passageway from the flue collar on the appliance or draft hood to the outside atmosphere for the purpose of removing the exhaust combustion flue gases.

The vent system consists of a vent or a chimney and vent connector that form an open and unobstructed passageway to the exterior.

A natural draft venting system removes flue or vent gases under negative static vent pressure entirely by natural draft. Natural draft systems utilize the stack effect of the chimney, as well as the density differences in the heated combustion gases, to create natural gas flow through the appliance. See Figure 3-7.

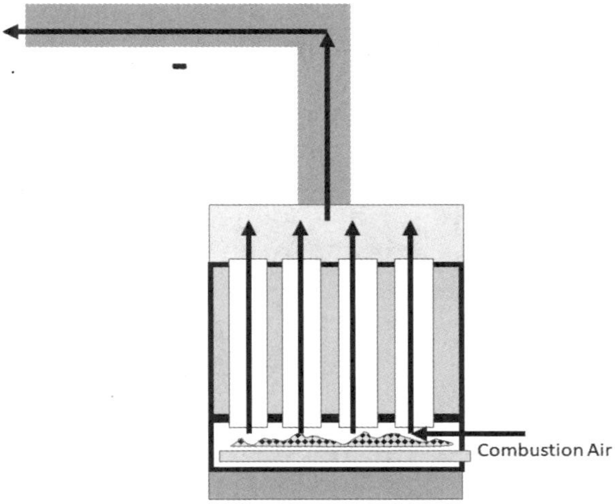

FIGURE 3-7 Natural Draft Appliance

Gas-fired appliances that are categorized for the purpose of vent selection are classified into the following four categories:

Category I. An appliance that operates with a negative vent static pressure and with a vent gas temperature that avoids excessive condensate production in the vent

Category II. An appliance that operates with a negative vent static pressure and with a vent gas temperature that is capable of causing excessive condensate production in the vent

Category III. An appliance that operates with a positive vent static pressure and with a vent gas temperature that avoids excessive condensate production in the vent

Category IV. An appliance that operates with a positive vent static pressure and with a vent gas temperature that is capable of causing excessive condensate production in the vent

Direct Venting

Building and appliance designs have resulted in many modern boilers and furnaces directly venting products of combustion to the exterior, eliminating the use of a conventional chimney.

A mechanical draft venting system uses either an induced draft, where a portion of the combustion gas path is under a negative static pressure, or a forced draft, where a portion of the gas path is under positive static pressure. This is accomplished with an induced draft fan or power venting fan.

In the induced draft venting system, a portion of the venting system uses an induced draft fan to remove the flue or vent gases by drawing them away from the appliance under negative static vent pressure (Figure 3-8).The fan can be either integral with the boiler or furnace in which it is mounted or separate and interlocked with the appliance's burner, the latter of which is known as a power venter.

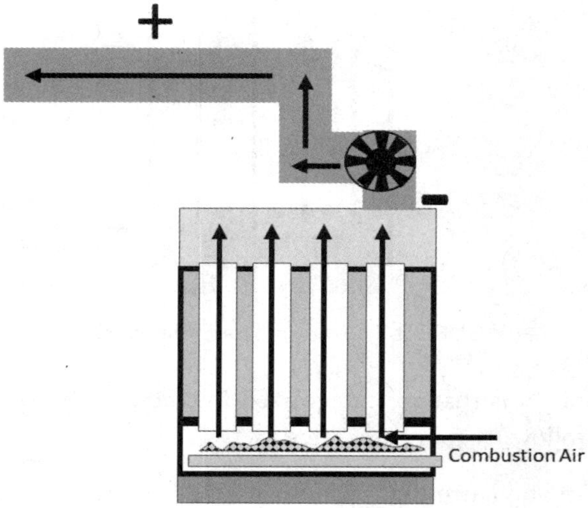

FIGURE 3-8 Induced Draft

The forced draft venting system utilizes a forced draft fan to "push" flue or vent gases through the vent system under positive static vent pressure (Figure 3-9).

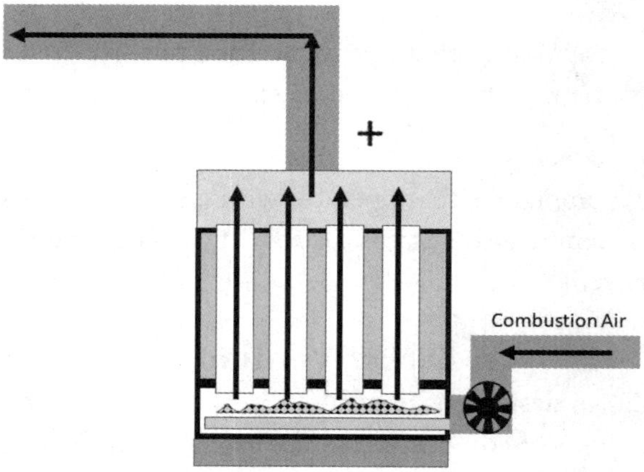

FIGURE 3-9 Forced Draft

It is important to note the allowed locations of direct vent exits through the exterior walls of a building envelope. Care must be taken to provide proper clearances from window openings, intake combustion air sources, and areas that may cause damage or injury (Figure 3-10).

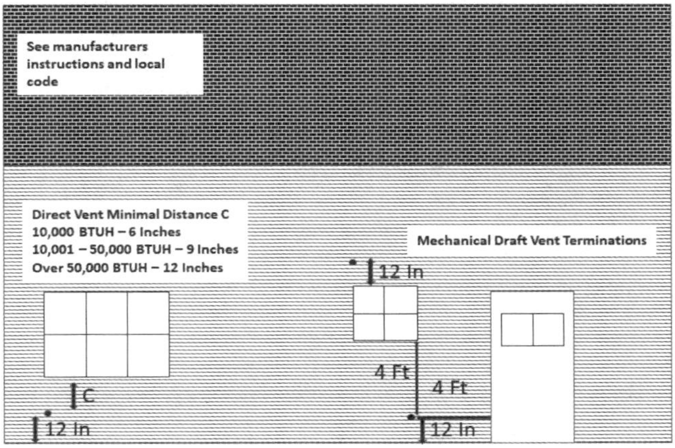

FIGURE 3-10 Termination Points from Building Openings for Direct Vent Appliances

Care must also be taken to make sure that the termination points of all combustion appliances are not positioned in a location where these emissions can be hazardous, and as well that the termination points are not in a location that would facilitate the reentry of these emissions into the structure.

Wind, air density differences, and air-exhausting appliances and fixtures can cause an air pressure imbalance between the inside and exterior of a building, a state known as *depressurization*. Depressurization combustion reentry occurs when combustion emissions leave the heating system through the system exhaust venting as intended but are drawn back into the structure through building openings and infiltration air accumulating over time.

Forensic Insight

Chimney Termination Point Problem

For combustion appliances utilizing chimneys (Figure 3-11), problems with indoor air quality can occur when emissions from heating systems discharge from the appliance as intended but reenter the structure. The chimney termination point should be considered when this issue occurs since this termination point can be the culprit, causing these emissions to strike the roofline and allow products of combustion to reenter the structure through air infiltration paths. A chimney that is below adjacent building rooflines or not high enough can be a contributor to this problem.

FIGURE 3-11 Chimney Termination Point Below Roofline

Practical Considerations Addressing the Installation of Heating Systems

Boiler and Furnace Placement

In deciding where to place a boiler or furnace, it is necessary not only to take into consideration fire safety, boiler or furnace sizing, but also the identification of each room's radiant or hot air input requirements under design conditions. It is necessary also that practical considerations be given to the installation of heating systems.

For example, because of budget constraints, many homes are built with unfinished rooms. So if possible, when deciding the output of the boiler or furnace, take into account potential plans to remodel and expand into initially unfinished attics, basements, or other areas.

For instance, say a house was initially built with an unfinished basement. Installation of the boiler in a catch pan plumbed to a suitable drain (Figure 3-12) greatly reduces the potential for building damage due to a water discharge or leak if the basement is remodled later on.

If these appliances are in a space such as an unfinished basement or other area that will be remodeled in the future, to prevent future water damage in the event of an equipment malfunction, the best time to make accommodations and install protective pans, as well as install discharge and condensate piping, is during the installation of these appliances.

Access for Maintenance and Replacement

Access for maintenance is another consideration in the placement of heating and cooling appliances. Boilers and furnaces must be maintained on an annual basis, and access to these appliances to allow this maintenance is a code requirement as well as a practical necessity. Maintenance will include cleaning, inspection, and sometimes major repairs such as the replacement of a combustion chamber or heat exchanger, which will require disassembly as well (Figure 3-13).

FIGURE 3-12 Boiler Installed in a Catch Pan Plumbed to a Suitable Drain

FIGURE 3-13 Boiler Clean Out Ports Inaccessible for Proper Maintenance

Also, boilers and furnaces should not be built into inaccessible building cavities or spaces, because these appliances will not outlive the building! At some time, the appliance will likely have to be replaced. Building demolition to allow the removal and reinstallation of a heating appliance is an unnecessary foreseeable burden.

Placement of Ductwork Diffusers and Baseboards

In addition to identifying the needed heat input requirements for each room, it is essential to carefully select where in the room the heating duct or baseboard is placed. The close proximity of a thermostat to a heat register or radiator will cause frequent "on-off" short cycling of heat in the room as well as areas of colder air.

Registers and radiant baseboards should not be installed near or adjacent to plumbing fixtures that may expose them to water or human waste, such as a radiant baseboard placed adjacent to a toilet (Figure 3-14) or shower stall. Over time, exposure to cleaning and other liquids will create corrosion and other unsightly damage that could have been avoided with proper planning.

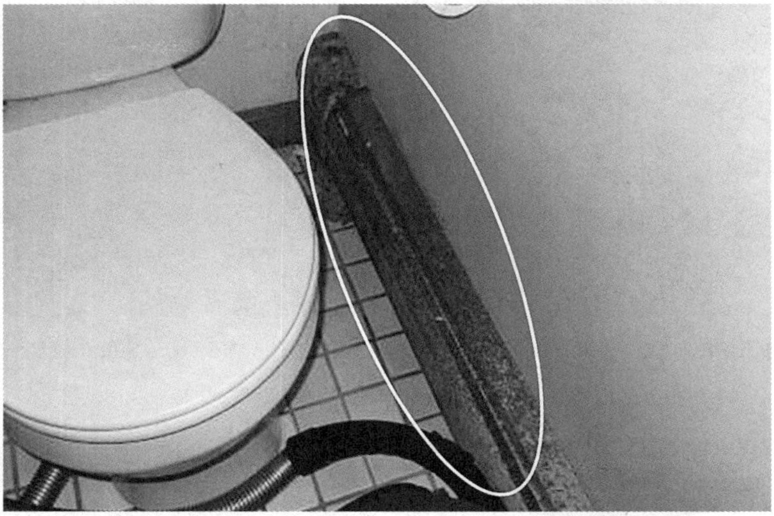

FIGURE 3-14 Poor Decision for Placement of Baseboard Element

Freezing

Consideration must be given to where appliances are placed to avoid damage due to freezing (Figure 3-15). And the potential for freezing and causing damage must also be considered in the installation of condensate and water-bearing piping in or passing through an unconditioned space or cavity.

FIGURE 3-15 Piping Located in Unheated Building Space Prone to Freezing

Unlike plumbing codes that address water systems for drinking and sanitary waste discharge, the code guidelines that govern hydronic heating system piping, at this time, do not directly address hydronic system freeze protection. The IECC, or International Energy Conservation Code, includes requirements for hydronic system piping to be insulated when it passes through an unconditioned space. However, this is intended to address energy (heat) losses to that cold unconditioned space This practice inadvertently improves the ability of the piping to resist or delay freezing, but it does not make it incapable of freezing.

References to freeze protection of appliances and water-bearing piping including hydronic heating system piping are occasionally provided in the manufacturers' installation instructions. Also the *I=B=R Guide to Residential Hydronic Heating* addresses the need for inhibited propylene glycol antifreeze protection of hydronic heating system piping susceptible to freezing.

References

International Building Code, 2009.

International Mechanical Code, 2012.

International Energy Conservation Code, 2015.

NFPA 31: Standard for the Installation of Oil-Burning Equipment, 2011.

NFPA 54/ANSI Z223.1: National Fuel Gas Code, 2012.

NFPA 211: Standard for Chimneys, Fireplaces, Vents, and Solid Fuel Burning Appliances, 2010.

CHAPTER 4

BOILER DESIGN

Boiler Construction

The heart of a steam, hydronic (forced hot water), or hydro-air heating system is the boiler. The boiler exchanges heat from the combustion of fuel oil, natural gas, or propane (or even electrical resistance) and transfers that heat to the water in the boiler's pressure vessel. Boilers are typically limited to an operating pressure of 30 psi.

The heat exchanger of the boiler may be made of cast iron, steel, or other material and is typically wrapped in insulation over which is placed a sheet metal jacket. Regardless of what the boiler's heat exchanger is made of, the basic nomenclature of the areas of the boiler is universally understood.

As seen in Figure 4-1, the combustion of fuel occurs in the boiler's *combustion chamber* (A), or *firebox*, and then hot combustion gasses pass through the boiler's heat exchanger *flue passageways* (B).

The boiler's heat exchanger can be built in several different ways. In the example in the figure, the boiler is built with hollow (water-filled) *cast-iron sections* joined together (C). Upon exiting the flue passageways, products of combustion then enter the *collector box* (D), which is connected to the discharge *venting pipe* (E) that leads the products of combustion out of the boiler and into a conventional chimney flue or mechanical venting system.

Typically, the firebox is lined with a high-temperature *refractory* (F) designed to protect the metallic components of the boiler from the high heat of the burner flame as well as act as a thermal reservoir allowing heat transfer to the boiler water after the burner has shut down.

The surfaces of the boiler in contact with products of combustion are known as the *firesides*, while those in contact with water or steam are referred to as the *watersides*.

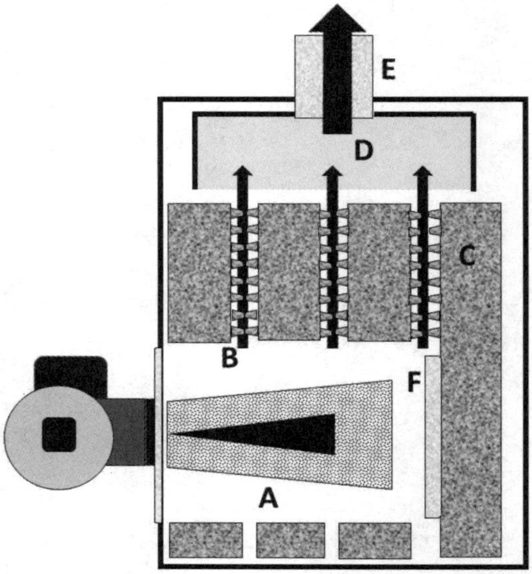

FIGURE 4-1 Boiler Cross Section Showing Flue Passageways

Boiler Configuration—Wet Base and Dry Base Boilers

In a *wet base boiler* (like the one seen in Figure 4-1), the boiler's flame is completely contained within a boiler section that is surrounded by water. Although this type of boiler utilizes an interior refractory inside the combustion chamber area, it does not rely on the integrity of this refractory to contain the burner flame and products of combustion to prevent boiler damage and product-of-combustion leakage.

Furthermore, with a water-filled portion of the heat exchanger surrounding the burner flame, installations adjacent to combustible surfaces will likely be less restrictive than alternative boiler designs where the burner flame is only contained and surrounded by an insulated firebox.

A boiler where the burner flame is only encapsulated by the steel and refractory-insulated firebox is known as a *dry base boiler*.

Cast iron is used in the majority of boilers in residential as well as light commercial applications and has been used for this purpose since the 1800s. In fact, many "antique" cast-iron boilers are still in service today in all types of heating systems, such as steam, gravity, and forced hot water systems. However, cast iron boiler design has evolved over time in steps to increase efficiency and ease of maintenance and installation.

Cast-Iron Sectional Boilers

Most modern cast-iron boilers are constructed by assembling any number of hollow cast-iron sections together (Figure 4-2). After assembly, products of combustion flow between these sections through flue passageways, giving off their heat to the water contained in the sections (Figure 4-3). As shown in the bisected photograph of a cast-iron boiler section in Figure 4-4, the sections are cast with internal bracings between the casting walls, strengthening the boiler section walls.

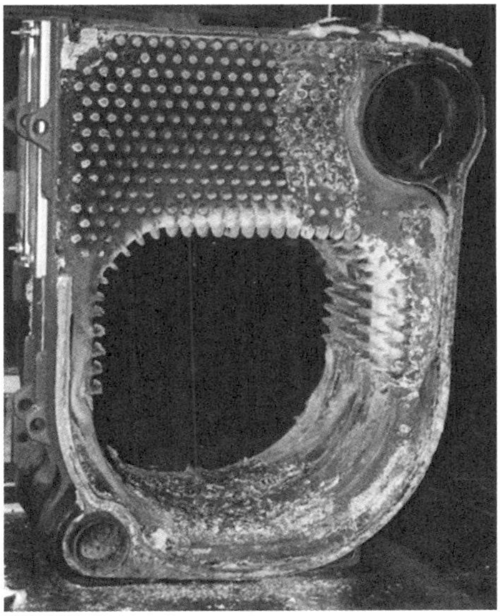

FIGURE 4-2 Cast Iron Boiler Sections Exposed

FIGURE 4-3 Flue Passageways Between Cast Iron Boiler Sections

The firesides of cast-iron boilers are also cast with several small protrusions, known as *heat pins* (Figure 4-5). The heat pins act to increase the heat transfer surface area of the boiler, increasing the efficiency of the boiler's heat transfer. Manufacturers design the pattern of the heat pin field to allow the passage of a cleaning "sword" or brush to dislodge soot during cleanings.

FIGURE 4-4 Bisected Cast Iron Boiler Section

FIGURE 4-5 Boiler Fireside Heat Pins

The watersides of these sections are filled completely with water for forced hot water systems or only partially filled for boilers supplying steam heating systems.

Cast-Iron Boiler Joining Methods

Boiler sections are held together by a threaded rod (also known as a draw rod, shown in Figure 4-6) affixed to the end sections, holding the sections together in compression. Each boiler section is sealed to the adjacent section by either of two means: a gasket or an internal, threadless fitting known as a push nipple.

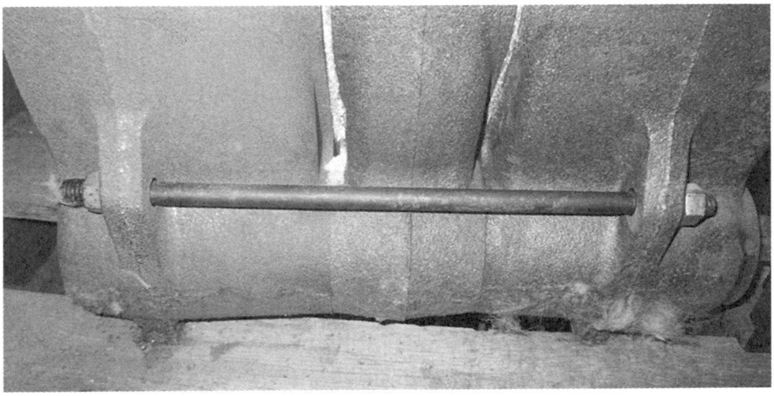

FIGURE 4-6 Draw Rod Holding Boiler Sections Together

Section Gaskets

As seen in Figure 4-7, one method of joining and creating a watertight seal between cast-iron boiler sections is to use a section gasket. This type of gasket is typically made from a resilient synthetic rubber product such as EPDM (ethylene propylene diene monomer).

FIGURE 4-7 Boiler Section Gasket

Boilers that use gasket seals between sections provide ease of assembly when constructed in the field. This is a common way to construct larger boilers, where the size and weight of each section dictates that the sections be brought to the machinery space individually.

Likewise, in the event that a section is damaged causing it to leak, the individual section can be easily removed and replaced since the section gaskets are easily removed and replaced.

Section Push Nipples

The push-nipple method of joining boiler sections creates a watertight–steam-tight seal by creating an interference fit between the boiler casting and the steel push-nipple fitting, as shown in Figure 4-8.

Typically, for smaller residential boilers the boiler heat exchanger is shipped completely assembled and ready for the addition of controls (or trim). For boilers that are assembled in the field, such as for larger commercial applications that require the sections be brought into the building individually, boilers constructed with push nipples require a larger threaded rod "assembly tool" to draw the sections together and press them together.

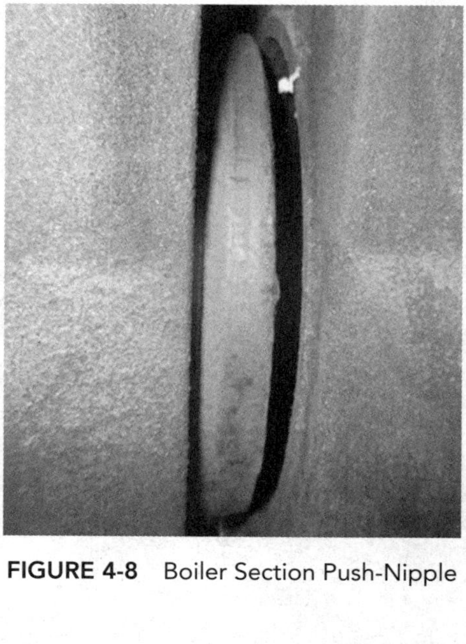

FIGURE 4-8 Boiler Section Push-Nipple

Section Sealant, Rope Gaskets, or Sealing Cord

Products of combustion are prevented from leaking out from between the sections with a rope gasket (aka sealing cord) or sealant placed around the perimeter of the boiler section firesides where each section meets (Figure 4-9). These seals can, over time, wear and allow products of combustion to leak from the sections and into the space between the cast-iron boiler block and boiler jacket (Figure 4-10).

FIGURE 4-9 Rope Gasket Sealing Combustion Chamber Area in Boiler Castings

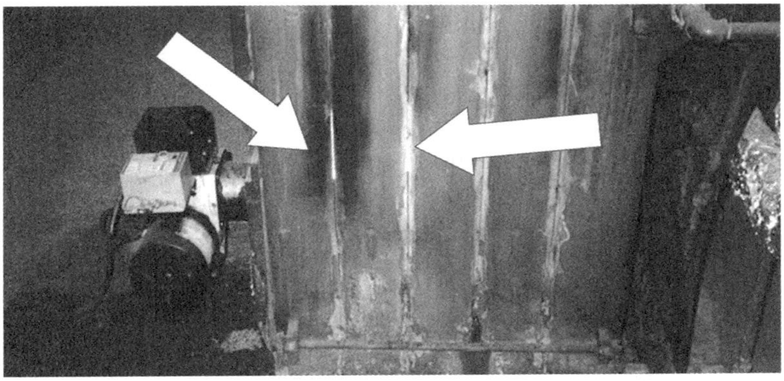

FIGURE 4-10 Product of Combustion Gas Leakage from Between Boiler Sections

During preventative maintenance, if products of combustion (soot or moisture) are observed on the insulation between the boiler block and outer jacket, this is an indication of a possible product-of-combustion gas leak, and it should initiate an inspection of any cleanout service door gaskets as well as the seals between boiler sections.

Carbon monoxide leakage or a combustion performance test of the burner that yields a low carbon dioxide reading from a combustion performance test may also be an indication that air is leaking into the firebox and/or that the section seals may be leaking.

Burner Type and Flame Geometry and Orientation Overview

Flame geometry is governed by the manner in which the fuel is being consumed.

For boilers using heating oil or gas, oil burners or conversion "power" burners combust fuel by using either an atomized spray of fuel oil or a flow of fuel gas and mixing it with combustion air. As seen in Figure 4-11, this results in a cone-shaped flame geometry that is usually horizontal and shaped by nozzle size, pump pressure, and airflow to conform to the dimensions of the appliance's combustion chamber.

FIGURE 4-11 Oil Burner Flame

For gas-fired appliances not utilizing a conversion power-burner, the typical burner arrangement, depicted in Figure 4-12, consists of a gas valve admitting natural gas or propane to a gas manifold. The manifold is outfitted with several orifice nozzles, each of which is connected to a burner tube that is also outfitted with a primary air adjustment.

FIGURE 4-12 Conventional Gas Burner

As Figure 4-12 shows, the flame geometry consists of multiple rows of burner flames firing in a vertical direction. Proper flame adjustment prevents the flames from striking any of the boiler's firebox components, which would cause soot production and flue passageway obstruction.

Boiler design and construction configurations are based on the type of fuel to be used as well as the type of burner the appliance will be outfitted with and the resulting flame geometry and dimensions. More details regarding burner design will be presented in Chapters 6 and 7.

Tankless Heat Exchangers and Gaskets

Domestic water can be heated using the large water mass and thermal inertia of larger cast-iron and steel boilers by installing a heat exchanger coil into the boiler. The coil heat exchanger, known as a *tankless hot water coil*, is constructed of a "finned" copper tube wound into a coil shape (Figure 4-13) and is also enhanced by the extra surface area provided by the addition of the finned surfaces on the outside wall of the coil.

As seen in Figure 4-14, the domestic tankless hot water coil is integrated into the boiler with the heat exchanger and immersed in the hotter boiler hydronic or steam water. Since the heating system side of the boiler is usually operated in the 12–15 psi range but limited to 30 psi and the domestic water operates at a higher pressure limited to 80 psi by code, a leak in the domestic water coil will cause water to run into the boiler. This will increase the water pressure in forced hot water systems, causing steam systems that are not filled entirely during normal operation to overfill and thus the entire system to become water filled.

FIGURE 4-13 Domestic Hot Water Coil

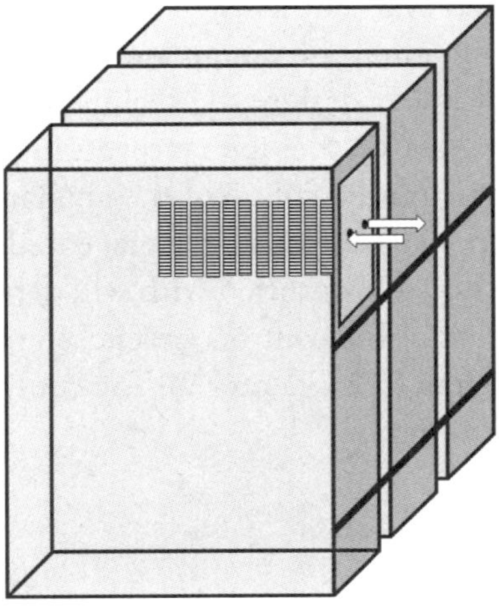

FIGURE 4-14 Placement of Domestic Hot Water Coil in Boiler Section

A relatively simple and inexpensive gasket replacement repair can fix the leaky coil. If the leak is left unchecked, the leaking water can result in extensive corrosion damage to the boiler, making repair and replacement of the coil more difficult or potentially not even possible due to the extent of corrosion damage, rendering the entire appliance in need of replacement (Figure 4-15).

FIGURE 4-15 Extensive Corrosion Damage Caused by Prolonged Tankless Hot Water Coil Leakage

Boiler Heat Exchanger Design Configurations

Boilers are designed and built in several styles and using several kinds of materials and methods. To ensure ease, success, and cost-effectiveness of maintenance, the better the technician knows how the appliance is built, the better he or she can access the various components of the appliance for inspection and replacement if needed.

The type of burner used also directs the design configuration of the boiler. Block design for an oil or gas power-conversion type of burner with a horizontally aimed flame firing into the combustion chamber is entirely different from the design for a standard gas-fired or induced draft fan-assisted burner.

Also, by better understanding the design and construction details and the vulnerabilities of each type of boiler design, the technician can more effectively focus on the susceptible components of the appliance during preventative maintenance intervals. By doing so, a catastrophic failure of the appliance, rendering it in need of emergency replacement or potentially causing damage to the building, can be avoided.

The controls that the boiler is outfitted with will depend on the type of heating *system* the boiler is connected to. The controls are chosen to conform to and allow the boiler to work with this system (see Chapter 9). The controls will govern the boiler's operational parameters regarding:

- Burner operation
- Boiler water temperature
- Boiler water level
- Boiler water or steam pressure

The design details of the boiler's construction should be completely understood by the servicing technician. This understanding should take into account observations of similar equipment failures experienced during service events.

Also, forensic examinations of past failed equipment can and should provide a valuable lesson when servicing *operational* equipment since past failures can and likely will manifest themselves in other failures of a similar design unless the servicing technician intervenes.

By better understanding the appliance's design and component materials' behavior, the servicing technician can focus on the vulnerable components of a boiler's construction, such as the water-steam connections between sections or the refractory and combustion gas seals, and will be able to access them through disassembly of the appliance during servicing.

Of course, boiler heat exchanger and control designs are continually evolving and changing. This means that across the entire population of current in-service boilers, the servicing technician will encounter boilers of vastly different ages, fuel sources, and heating systems. However, the basis of their operation remains the same.

Most of a heating service technician's time is not devoted to installing new, current, state-of-the-art boilers but rather keeping existing systems in operation. As such, a familiarity of all types of boiler designs is needed for effective repair and preventative maintenance. Following are typical boiler design configurations and heat exchanger constructions that the service technician may encounter.

Vertical Fire-Tube Cast-Iron Boiler ("Snowman")

The evolution of cast iron in boiler construction began at the turn of the twentieth century when the fuel of choice was coal. One popular and resilient boiler is the coal-fired, vertical fire-tube cast-iron boiler, as depicted in Figure 4-16. Like the boiler in the figure, virtually all these "antique" boilers still in use today have been converted from burning coal to burning gas or heating oil by the retrofitting of a gas or oil burner.

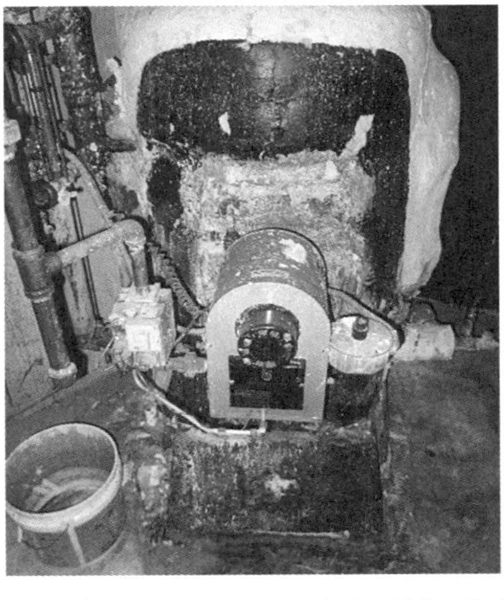

FIGURE 4-16 100-Year-Old Cast Iron Vertical Fire Tube Boiler Still in Use Today

The boiler pictured in Figure 4-16 was constructed of hollow, water-filled cast-iron sections. The middle section has vertical "fire tubes" (Figure 4-17) that allow the passage of products of combustion from the combustion chamber below.

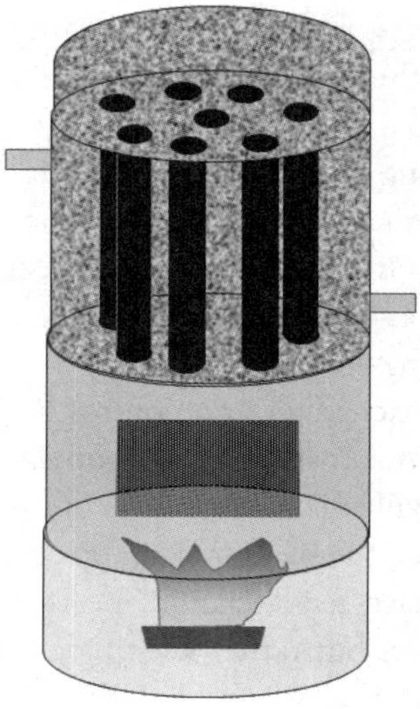

FIGURE 4-17 Illustration of Cast Iron Vertical Fire Tube Boiler Design

As seen in Figure 4-18, initial overheating protection was provided by a "fusible plug" that consisted of a hollowed-out tin-filled center, located at the bottom of the water-filled section and at the top of the boiler's combustion chamber area. In the event that the boiler would overheat because the water in the boiler was too low, the tin in the plug would melt, causing whatever water or steam was left in the boiler to pass out of the newly created opening above the combustion chamber, extinguishing the coal fuel in the chamber. In modern configurations newer safety controls are typically added to these boilers, such as low-water cutoff switches in steam heating systems.

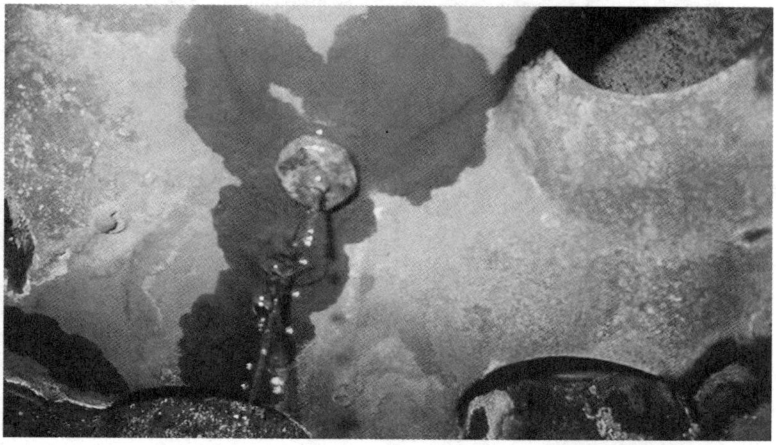

FIGURE 4-18 Older Boiler's Fusible Plug Low-Water Cutoff Protection Device

Extreme care must be exercised when repairing and especially when removing these and similar older units since they are insulated on the exterior with asbestos. If handled improperly, loose asbestos can be hazardous to the health of technicians and occupants of the home.

Power Burner–Fired (Oil or Gas Conversion–Fired Boiler), Cast-Iron Sectional Boiler—Wet Base

One of the most common and popular boiler designs for oil-fired or power burner–fired installations is the wet base, cast-iron sectional boiler, shown in Figure 4-19. Because the burner flame is completely encircled by the boiler's water-filled heat exchanger, as illustrated in Figure 4-20, the boiler can usually be installed adjacent to combustible surfaces more often than alternative boiler designs and provides greater thermal efficiency.

FIGURE 4-19 Wet Base Cast Iron Sectional Boiler

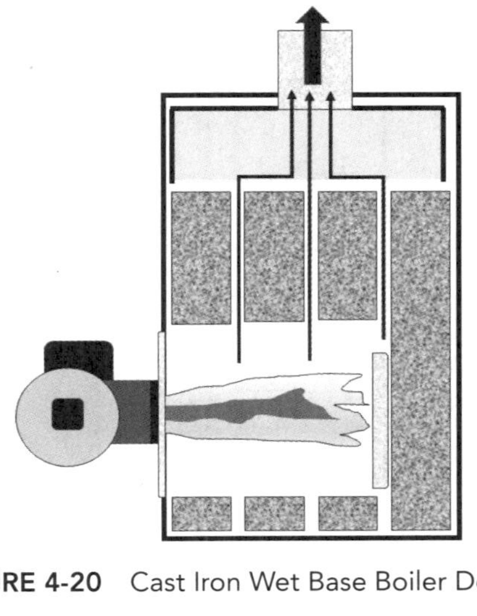

FIGURE 4-20 Cast Iron Wet Base Boiler Design

The boiler's combustion chamber is lined with a ceramic fiber refractory material, as seen in Figure 4-21, that serves as a heat sink as well as to protect the cast-iron boiler section from the direct heat of the burner flame, and sections are joined with either a steel push nipple or a synthetic rubber EPDM type of gasket.

FIGURE 4-21 Cast Iron Boiler Showing Combustion Chamber Refractory in Place

Steel Fire-Tube (Oil or Gas Conversion–Fired) Boiler—Dry Base

Another common boiler design is the steel fire-tube dry base boiler, shown in Figure 4-22.

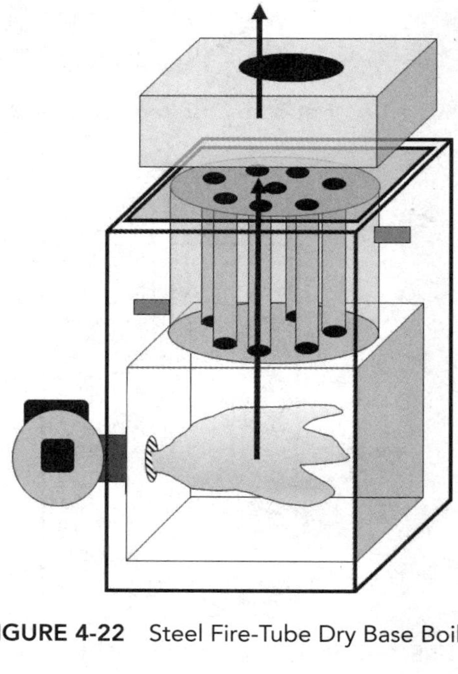

FIGURE 4-22 Steel Fire-Tube Dry Base Boiler

In this type of boiler, as shown in Figure 4-23, the heat exchanger is constructed with "fire tubes," or flue passageways, that pass through a water-filled boiler pressure vessel where heat from the products of combustion is transferred to the water within. The heat exchanger fire tubes in steel fire-tube boilers can be oriented vertically or horizontally. The heat exchanger in Figure 4-23 is placed on top of or adjacent to a steel, refractory-lined firebox (Figure 4-24) that receives its products of combustion from a horizontally oriented oil- or gas-fired power burner.

FIGURE 4-23 Steel Boiler Heat Exchanger

FIGURE 4-24 Steel Refractory-Lined Combustion Chamber Firebox

The entire firebox–heat exchanger assembly is insulated with fiberglass insulation and an outer sheet metal jacket (Figure 4-25).

FIGURE 4-25 Disassembled Dry Base Boiler Showing the Heat Exchanger and Dry Base Combustion Chamber

The design of this boiler configuration makes the integrity of the firebox's combustion chamber refractory vital to the boiler's prolonged operation since the heat of the burner flame can melt the steel wall of the boiler's firebox. This will cause products of combustion from the burner to leave the appliance, as seen in Figure 4-26. The figure shows damage to a steel dry base boiler's firebox due to damage to the combustion chamber refractory caused by a service technician's cleaning tools accidentally striking the refractory when the boiler's flue passageways were cleaned during servicing.

FIGURE 4-26 Cleaning Brush Damage to Combustion Chamber Refractory

Another unique steel boiler design incorporates a spiral heat exchanger. This type of boiler, shown in Figure 4-27, has a low water volume that increases efficiency.

FIGURE 4-27 Steel Spiral Boiler Heat Exchanger (Courtesy of Energy Kinetics)

Power Burner–Fired (Oil or Gas Conversion–Fired Boiler), Cast-Iron Sectional Boiler—Dry Base

An earlier configuration of the cast-iron sectional boiler similar to the steel fire-tube dry base boiler design is the cast-iron sectional dry base boiler for use with oil or gas power-conversion burners. As seen in Figure 4-28, products of combustion are delivered to the boiler's combustion chamber from an oil or conversion gas burner and pass through a cast-iron sectional boiler heat exchanger placed on top of this firebox and then discharged from the building through a traditional chimney or power vent system.

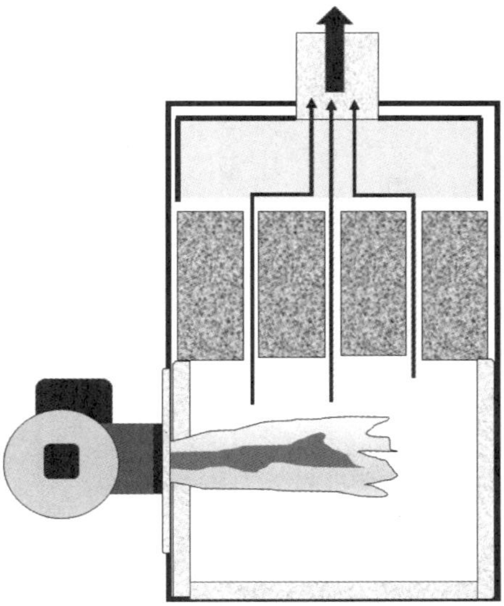

FIGURE 4-28 Cast Iron Oil or Power Burner-Fired Dry Base Boiler

Like the design for the steel boiler, the combustion chamber design in the dry base cast iron boiler is one of only a refractory-lined steel firebox that contains the burner flame and its products of combustion. The boiler's combustion chamber is accessed for inspection and servicing by removing the burner and its front mounting plate, as seen in Figure 4-29.

FIGURE 4-29 Burner Mounting Plate Removed, Revealing Inner Combustion Chamber

Cast-Iron, Gas-Fired Boiler—Dry Base

The standard design for smaller residential gas-fired boilers is the cast-iron dry base boiler, depicted in Figures 4-30 and 4-31. This configuration can also be augmented with an induced draft blower in higher-efficiency models to assist in combustion gas flow.

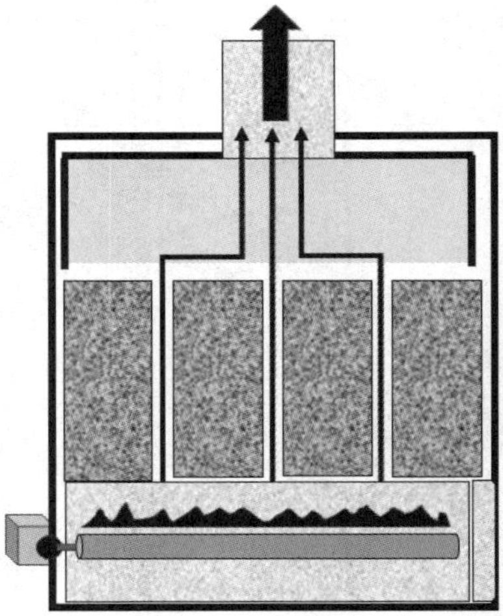

FIGURE 4-30 Gas-Fired Dry Base Cast Iron Boiler

Cast-iron sections are joined with either push-nipple or gasket connections, and the sections are insulated with fiberglass insulation over which the boiler's sheet metal jacket is fastened.

FIGURE 4-31 Gas-Fired Boiler

Combustion of fuel occurs in the boiler's combustion chamber, which consists of several gas burner tubes that are supplied with propane or natural gas through a typical gas valve–orifice system. The walls of the combustion chamber are lined with combustion chamber refractory panels, typically made of a ceramic fiber material, as shown in Figure 4-32.

FIGURE 4-32 Combustion Chamber Refractory Panels in Gas-Fired Boiler

Required preventative maintenance includes assessment of the gas burners, refractory, and heat exchanger for signs of wear, deterioration, or leaks. Care should be exercised to focus on the condition of the boiler's combustion refractory panels, especially the contact area between the heat exchanger and the combustion chamber refractory, to look for signs of product-of-combustion gas leakage.

Cast-Iron, Gas-Fired, Horizontal Sectional Boiler—Dry Base

Cast-iron boilers are composed of a number of cast-iron sections "stacked" together. Typically, boiler sections are assembled with each section being vertically oriented; however, a horizontally oriented boiler section design (known as a "pancake boiler") has been used in the past and is depicted in Figures 4-33 and 4-34. This boiler design accommodates gas combustion with the typical gas valve–burner tube design. Like all modern cast iron sectional boilers, the horizontally assembled boiler sections are insulated and encapsulated with fiberglass and a sheet metal jacket.

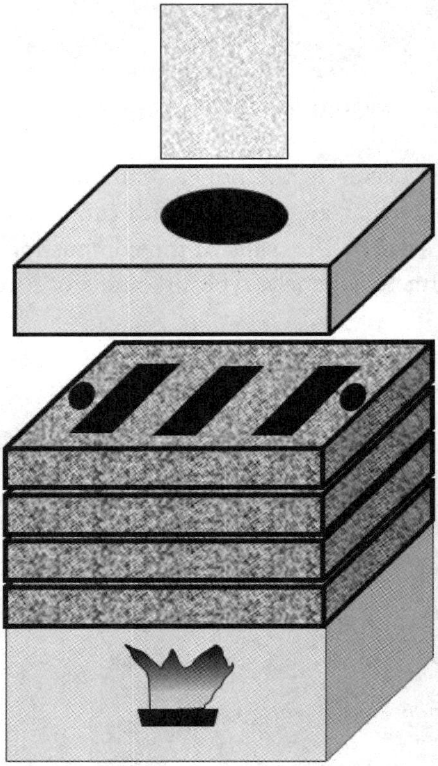

FIGURE 4-33 Dry Base, Cast-Iron, Gas-Fired, Horizontal Sectional Boiler

FIGURE 4-34 Horizontal Sectional Cast Iron Boiler Sections

Copper Tube–Aluminum-Finned, Low-Water-Volume, Gas-Fired Boiler—Dry Base

Building a boiler to contain a lower volume of water has several advantages. It may be built using an alternative construction method other than cast-iron sections, and it saves space. Another advantage is that it achieves a higher operating efficiency. A low-water-mass boiler also responds to demands for heat more quickly since there is a smaller amount of water to heat from the initiation of a call-for-heat cycle.

One heat exchanger design for the low-water-volume type of boiler utilizes a heat exchanger made of copper tubes outfitted with aluminum fins, as seen in Figure 4-35.

FIGURE 4-35 Copper Tube–Aluminum-Finned Boiler Heat Exchanger

In the header type of copper tube–aluminum-finned heat exchanger, which is also popular in swimming pool heater applications, the copper tubes are connected to a common header, one on each side of the inlet and outlet, and typically the connection between the copper tube and header is maintained with press fit or gasket seal (Figure 4-36).

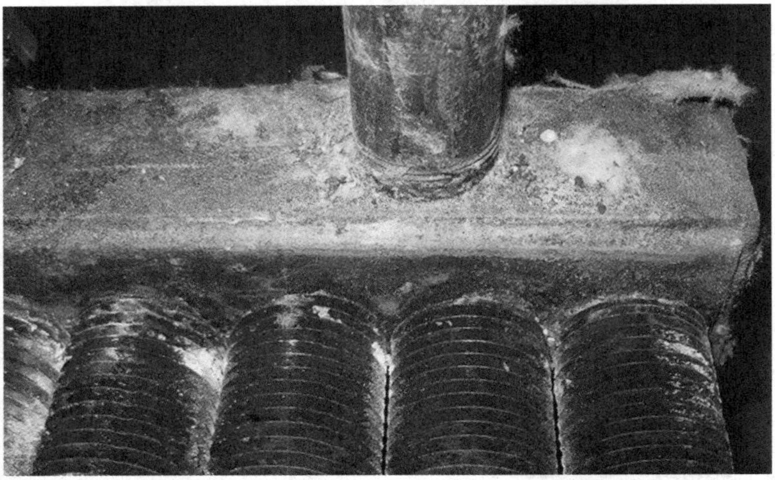

FIGURE 4-36 Copper Tube–Aluminum-Fin Tube Connection to Header

The gas path flow, as seen in Figure 4-37, leads from the gas burner system directly over the heat exchanger aluminum fins in a "one-pass" combustion gas path flow.

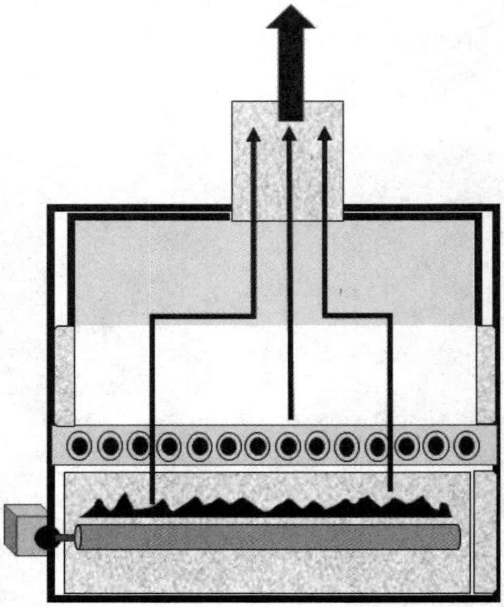

FIGURE 4-37 Gas Path Through Copper Tube–Aluminum-Fin Heat Exchanger

Coiled Copper Tube–Aluminum-Finned, Low-Water-Volume, Gas-Fired Boiler

Another copper tube–aluminum-finned heat exchanger design incorporates a coiled copper tube, as seen in Figure 4-38. In this design, a vertically fired forced draft gas burner fires directly into the coil-lined combustion chamber "dome," as seen in Figure 4-39 and illustrated in Figure 4-40.

FIGURE 4-38 Copper Tube–Aluminum-Fin Coiled Heat Exchanger

FIGURE 4-39 Coiled Heat Exchanger Combustion Chamber Dome

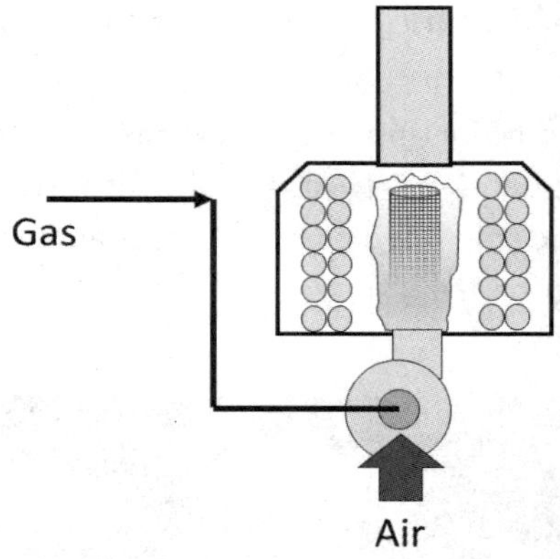

FIGURE 4-40 Illustration of Copper Tube–Aluminum-Fin Coiled Heat Exchanger

Spiral, Low-Water-Volume, Stainless-Steel, Gas-Fired, Condensing Boiler

One of the advantages of gas-fired, low-water-volume boilers is the decreased size of the boiler's heat exchanger, as well as its lower weight. See Figure 4-41. This has resulted in boilers with a smaller footprint, and it also permits the location of installation to be moved from the floor to the wall, creating more flexibility in areas where the appliance can be installed as well as easier access to the appliance.

FIGURE 4-41 Wall-Mounted Condensing Boiler

The decrease in size of these appliances is also enabled by efficiencies of heat transfer in the heat exchanger, resulting in lower exhaust temperatures. This is accomplished by greater heat extraction during the combustion process.

The amount of heat extraction from the products of combustion has increased to the point where the combustion gases condense (become liquid), requiring accommodations for condensate removal as well as other operational controls. The basic flow paths of a condensing boiler are depicted in Figure 4-42.

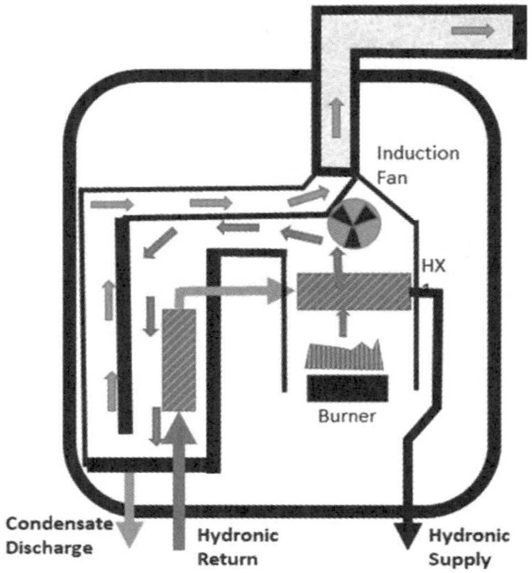

FIGURE 4-42 Condensing Boiler Flow Paths

Aside from the inherent electronic and electromechanical controls that govern the sequences of operation of these types of boilers, the most common design of the heat exchanger in these types of units is the low-mass, spiral, stainless-steel design, as depicted in Figures 4-43 and 4-44.

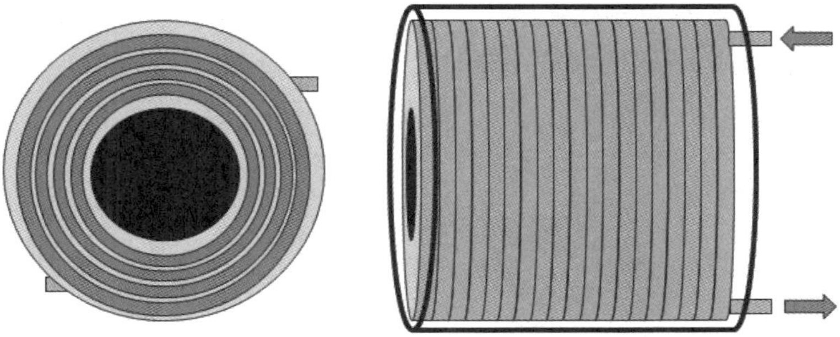

FIGURE 4-43 Illustration of Spiral Low-Mass Heat Exchanger

FIGURE 4-44 Fireside of Spiral Low-Mass Heat Exchanger

A stainless-steel heat exchanger provides resistance from the corrosive effects of combustion gas condensate, and the spiral tubular design of the heat exchanger results in a low-water-mass, thermally efficient heat exchanger, allowing the boiler to achieve operational temperatures and thermal efficiency more quickly. The burner fires directly into the middle of the heat exchanger (Figure 4-45).

FIGURE 4-45 Heat Exchanger and Burner from Low-Mass Condensing Boiler

Cast Aluminum

Another heat exchanger construction method used in high-efficiency gas boiler design is the single-piece cast-aluminum heat exchanger, as seen in Figure 4-46. This boiler design makes use of a low-water-volume and high-heat-conductivity heat exchanger to deliver higher thermal efficiency. As with most boilers of a condensing high-efficiency design, products of combustion are mechanically draft induced.

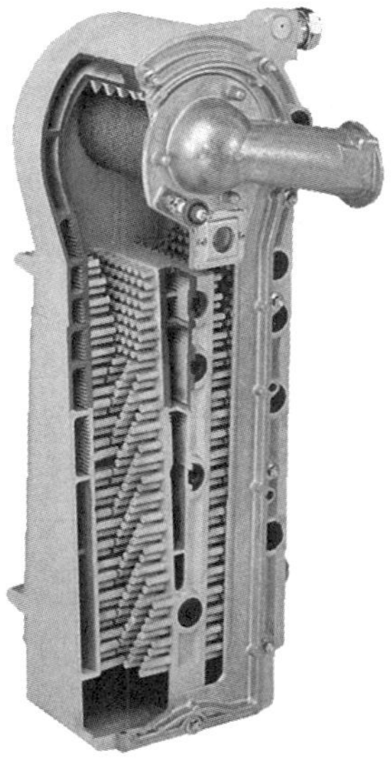

FIGURE 4-46 Aluminum Heat Exchanger (Courtesy of Weil McLain)

FORENSIC INSIGHT

Recognizing Overheating Damage to Boilers and Their Heat Exchangers

When a boiler's heat exchanger overheats, it can create great damage to a boiler of any service, as can overheating caused by hot products of combustion that leak. During preventative maintenance, impending catastrophic damage can sometimes be detected and caught in time to prevent further damage and perhaps save the appliance from needing to be replaced.

Damage caused by overheating can result from several conditions:

- Failure of the low-water cutoff switch
- Obstruction or closure of the water feeder valve
- Malfunction of the aquastat

Escaping hot combustion gases will also cause overheating damage to boilers, in turn causing overheating damage to components that were not designed to experience temperatures that these hot combustion gases impart to them.

Table 4-1 presents common materials that are used to construct boilers and lists their upper temperature limits in descending order. Knowing the upper temperature limits of these common boiler and furnace construction materials can provide insight about problems with an appliance's operation and installation conditions.

TABLE 4-1 Temperature Failure for Common Heating System Materials, °F

Combustion flame range: 3350°–3800°
Iron (melting point): 2800°
Steel (melting point): 2500°
Ceramic fiber (upper limit): 2100°–3000°
Copper (melting point): 1984°
Ferric oxide begins to form: 1600°
Fiberglass (melting point): 1400°
Aluminum (melting point): 1221°
Zinc (melting point): 787°
Tin (melting point): 449.5°
Enamel paint (scorches): 350°
Normally operating boiler water and steam temperature: 212°–220°

Heat Damage to Metallic Components

Heat-Induced Oxides

A deep red or a "black-blue" discoloration that appears on cast-iron or steel boiler or furnace firesides is an indication of overheating damage (Figure 4-47). The discoloration is caused by "heat-induced oxides," which are a combination of ferric and ferrous oxides. The oxides begin to form when the cast iron or steel approaches 1600°F. The damage is an indication of a loss of water in the boiler.

FIGURE 4-47 Boiler's Combustion Chamber Area—When Viewed in the Field, a Rust-Red Color as Well as a Black-Blue Color Indicate Overheating

Similar heat-induced damage to copper tube–aluminum-finned heat exchangers also results from low-water fired boilers and is identified by the consumption of the aluminum fins and a darker reddish discoloration of the copper components of the heat exchanger.

Cracks

In addition to heat-induced oxides, another sign of overheating damage to boiler and furnace heat exchangers is crack formation on the firesides (Figure 4-48). The cracks are linear in shape, and they should not be confused with holes. Cracks caused by thermal shock, however, can result without any easily visible heat-induced oxide formation.

FIGURE 4-48 Overheated and Cracked Steel Boiler's Heat Exchanger

Flame Impingement

For non-water-bearing components of a boiler or furnace heat exchanger such as the steel of a dry base boiler's firebox or furnace's heat exchanger under the combustion chamber refractory, direct and prolonged contact with hot product-of-combustion gases from flames will result in overheating and the steel melting (Figure 4-49). Typically, these areas are in direct contact and communication with the combustion chamber and burner of the appliance and often are enhanced by a loss of refractory material lining the firebox.

Melting of Nonmetallic Parts

The nonmetallic components of a combustion appliance can also be damaged by overheating. As seen in Table 4-1, these components have a lower melting point than the metallic components of the appliance's heat exchanger.

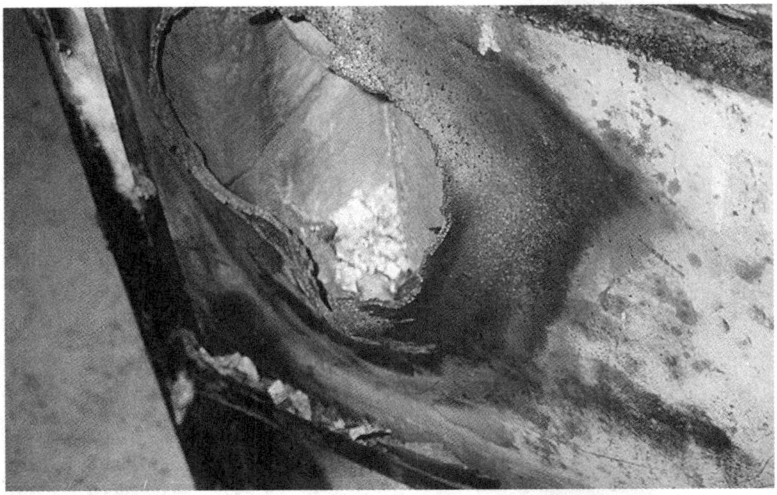

FIGURE 4-49 Flame Impingement Damage to Boiler's Steel Firebox

One example of overheating damage of nonmetallic components is melted or heat-damaged fiberglass insulation. Fiberglass is used to insulate the space between the heat exchanger and outer jacket and has a softening temperature of about 1400°. Abnormally high temperatures in the heat exchanger may eventually spread to the outer regions of the appliance, causing the fiberglass insulation to overheat and change color. When overheated, fiberglass insulation material, which is initially yellow, will characteristically turn white or even show charring (Figure 4-50).

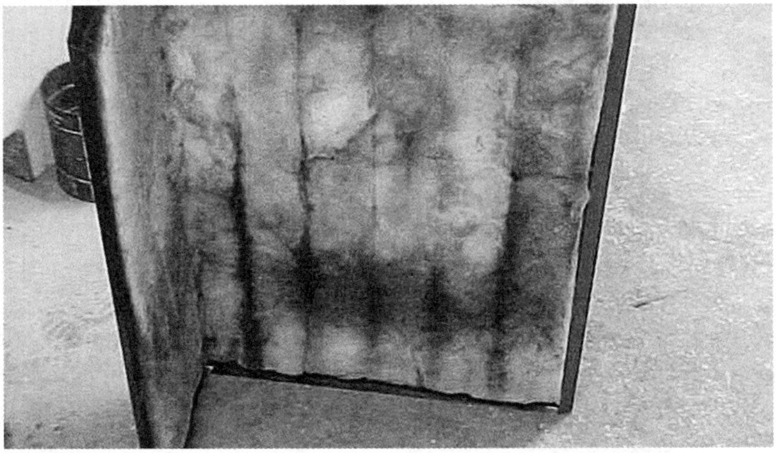

FIGURE 4-50 Overheated Boiler Insulation between Heat Exchanger and Outer Jacket

Overheating damage to insulation can also be concentrated in a smaller area and can be associated with product-of-combustion gas leakage into the insulated space between the heat exchanger–firebox assembly and sheet metal jacket. Just as overheating or combustion gas leakage causes insulation damage, oftentimes this damage extends to the outer jacket as well (Figure 4-51).

FIGURE 4-51 Boiler Jacket Paint Scorching

The gaskets in boilers are also subject to damage caused by overheating. As seen in Figure 4-52, an overheated boiler caused the synthetic EPDM rubber gaskets to melt and leak, rendering the boiler in need of either gasket renewal or total replacement.

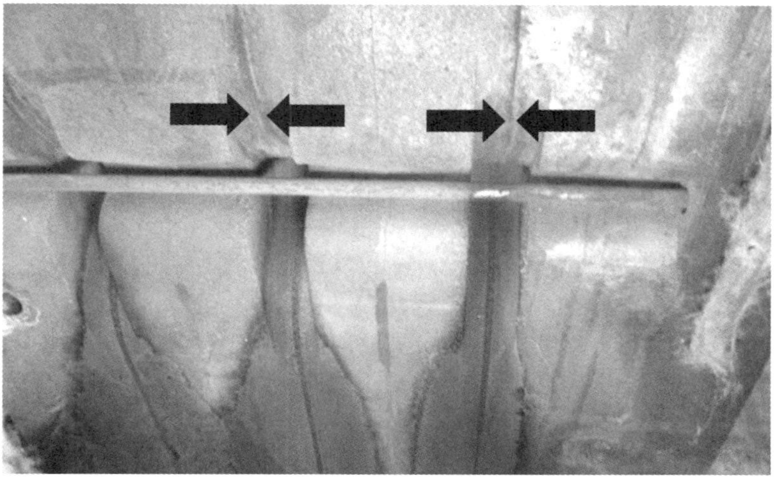

FIGURE 4-52 Overheated and Leaking Boiler Gaskets Between Sections

MAINTENANCE TIP!

Looking for Leaks

As part of an annual preventative maintenance program, aside from burner maintenance, examination of the boiler should include an assessment of any potential leaking. An unaddressed and prolonged leak from a boiler's heat exchanger will induce corrosion at the point of leakage as well as potentially affect boiler water chemistry and thus affect the boiler's life.

Because of the close tolerances of the boiler's heat exchanger, the heat exchanger is difficult to access. Still, common areas of leakage can be targeted when performing this preventative maintenance check.

Examples of common leaking locations include gaskets and push nipples.

Leaking Gaskets

Routine inspection of the gasket areas should be performed to identify any leaks that may have developed. For boilers outfitted with tankless hot water coils or those utilizing gasket seals in their construction, these areas can be easily accessed for inspection of potential leaks by removing the obstructing sheet metal jacket panel. Corrosion and hardness minerals (Figure 4-53) in these areas are an indication of leakage from these components' connections even if the boiler is not presently leaking.

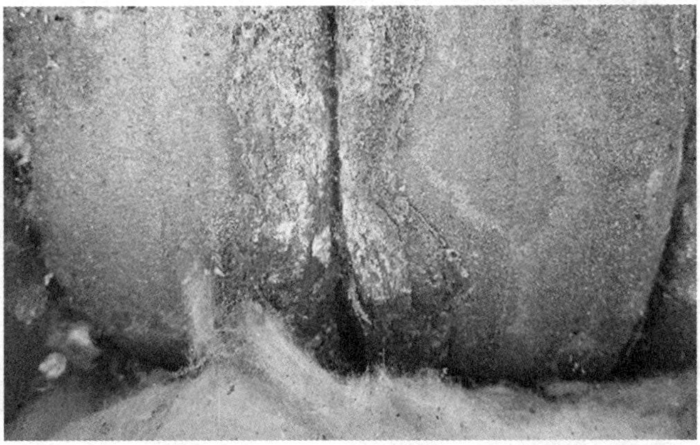

FIGURE 4-53 Examination of Gasket Area between Cast Iron Boiler Sections

As seen in Figure 4-54, if left uncorrected, corrosion can progress to the point where repair is more difficult if not impossible.

FIGURE 4-54 Examination of Boiler Section Gasket

Leaking Push Nipples

The push nipples in the boiler heat exchanger assembly also deserve inspection during preventative maintenance procedures or times when the boiler is suspected of leaking. Prolonged leakage can only worsen in time and cause the push nipples to fail. Figure 4-55 is a case in point. The steel draw rod in the figure held the cast-iron sections together which themselves were joined by a steel push nipple. The leakage occurred over such a long time period that it caused the draw rod to break apart, causing a water flood in the home.

FIGURE 4-55 Leaking Boiler Push Nipple

A leaking boiler is a boiler that needs immediate attention, and the damage will only get worse in time. Furthermore, depending on the location and severity of the leak, water damage to gas burners (as seen in Figure 4-56), the combustion chamber refractory, and boiler seals can disturb proper burner operation and cause burner malfunctions of a dangerous nature.

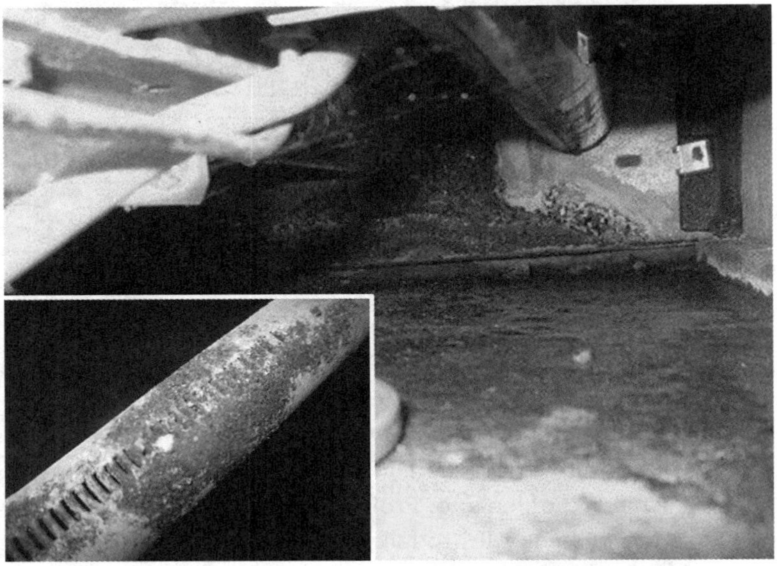

FIGURE 4-56 Water Leakage into Boiler's Combustion Chamber Damaging Refractory and Corroding Burner Tubes

TABLE OF BOILER HEAT EXCHANGER CONFIGURATIONS

Configuration	Example
Cast iron, vertical fire tube, dry base	Figure 4-17
Cast iron, power burner–fired, sectional, wet base	Figure 4-1
Cast iron, power burner–fired, sectional, dry base	Figure 4-28
Cast iron, gas-fired, horizontal sectional, dry base	Figure 4-33
Cast iron, gas-fired, dry base	Figure 4-30
Steel, vertical fire tube, dry base	Figure 4-22
Copper tube–aluminum-finned, low-water-volume, gas-fired, dry base	Figure 4-37
Spiral, low-water-volume, aluminum or stainless steel	Figure 4-43
Coiled copper tube–aluminum-finned, low-water-volume, gas-fired	Figure 4-40

CHAPTER 5

RESIDENTIAL FURNACE DESIGN

Warm Air System Types

As seen in Chapter 2, there are two ways to provide heated air directly to the interior of a structure: natural (gravity) hot air circulation and forced hot air circulation.

Natural (Gravity) Hot Air Circulation

Natural, or gravity, hot air circulation is a means of heat transfer where air is heated and circulated solely due to the natural air density differences in heated and unheated air. No matter if the heating appliance is a floor or gravity hot furnace, the circulation of heated air is accomplished by this means using a heat exchanger in the furnace.

The cycle of warm air being replaced by cooler air begins when fuel is burned on the primary side of the heat exchanger. The heat exchanger absorbs the heat of combustion and then transfers this heat to the secondary side of the heat exchanger. This secondary air is heated by direct contact with the secondary, or process, side of the heat exchanger. From here, the heated and less dense air rises away from the heat exchanger and is replaced by cooler incoming air.

Forced Hot Air Circulation

A more controlled means of hot air heating can be obtained by the forced circulation of heated air passing over a heat exchanger. In this method, airflow is mechanically controlled across the heat exchanger by means of a blower fan and a series of supply and return air ducts.

Furnace Designs, Natural (Gravity) Hot Air

Gravity hot air furnaces consist of a central heat exchanger that is enclosed in the body of the furnace cabinet. Typically, these older heat exchangers are made of cast iron or steel.

Air is supplied to and returned from the various rooms being heated by a large number of ductwork segments, causing the furnace to resemble and hence be known as an "octopus" furnace, as seen in Figure 5-1 (and also described earlier in Chapter 2).

FIGURE 5-1 Gravity Hot Air Furnace (Courtesy of "The Old House Guy")

Floor or Wall Furnace

For smaller houses or for single-room heating, another natural circulation gravity heating appliance is often used—a furnace set in the floor (Figure 5-2) or wall. In this type of appliance, a heat exchanger is heated by the combustion of fuel on the process side. The heat exchanger in the floor furnace is positioned within the furnace cabinet and covered by a floor grate.

FIGURE 5-2 Floor Furnace with Grate Removed Showing the Heat Exchanger

Furnace Designs, Forced Hot Air

The ability to control heat flow in hot air systems was realized by the implementation of forced hot air furnace construction. In this type of furnace, airflow is maintained over the outside surface of the heat exchanger by a blower fan and delivered to the various rooms through a system of ductwork. Air is returned from the various rooms through a separate series of return ductwork.

Airflow Direction

Upflow Furnace

The furnace configuration is chosen based upon the location of the ductwork connection in the mechanical room where the furnace will be installed. For ductwork located over the furnace, such as would be the case for an appliance installed in the basement, an upflow furnace is used (Figure 5-3).

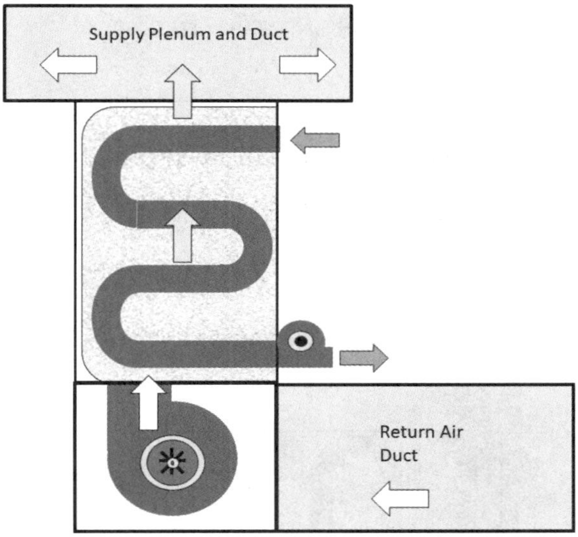

FIGURE 5-3 Upflow Furnace

Downflow Furnace

When the distribution supply and return ductwork is located below the furnace, such as in mobile homes, a downflow, or a "highboy," furnace is installed (Figure 5-4).

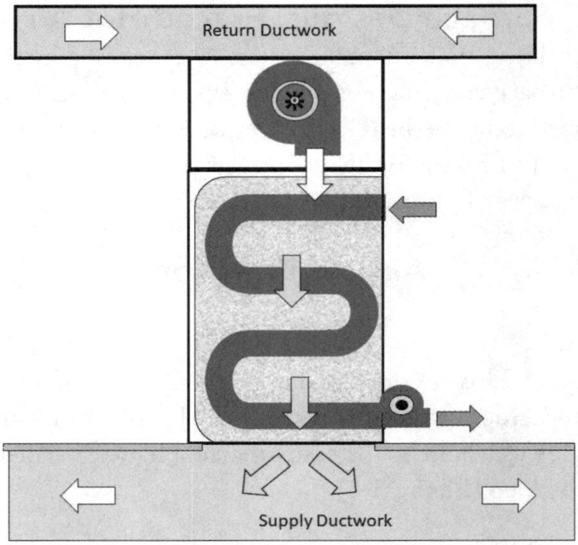

FIGURE 5-4 Downflow Furnace

Horizontal

Many furnaces as well as air handlers (including hydro air fan cabinets) are installed in small building spaces such as attics, eaves, or crawl spaces. Installations in these spaces are facilitated by the use of horizontally oriented furnaces or fan cabinets, as depicted in Figure 5-5.

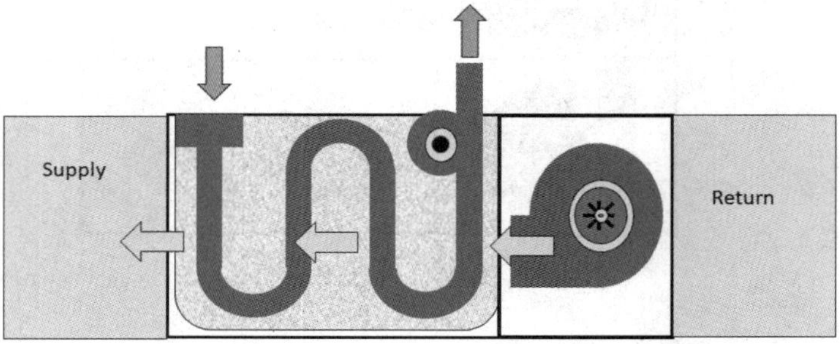

FIGURE 5-5 Horizontal Furnace

Energy Sources

The energy source for forced hot air appliances varies depending on local availability as well as the personal preference of the homeowner.

Local fuel providers of tank-based systems, such as propane or heating oil companies, may have difficulty in accessing rural roads during harsh winter months. This may influence the size and number of fuel tanks installed to ensure service without the fear of running out of fuel due to inaccessibility.

Energy sources provided by utilities, such as natural gas or electricity, offer the reassurance that an energy shortage due to property inaccessibility is not an issue.

The cost of various energy sources in various parts of the country will also impact the choice of fuel and the heating system design.

Electric Furnaces and Heat Pumps

Noncombustion energy sources of forced hot air systems are available and offer the convenience of no fuel lines, tanks, or combustion exhaust systems being needed.

Electric furnaces (Figure 5-6) accomplish heating with the use of electric resistance heaters within the "furnace" cabinet over which conditioned system air flows. Electric resistance heating elements are often energized in stages by sequencing controls that incrementally energize the heating elements in response to the time the thermostat calls for heat.

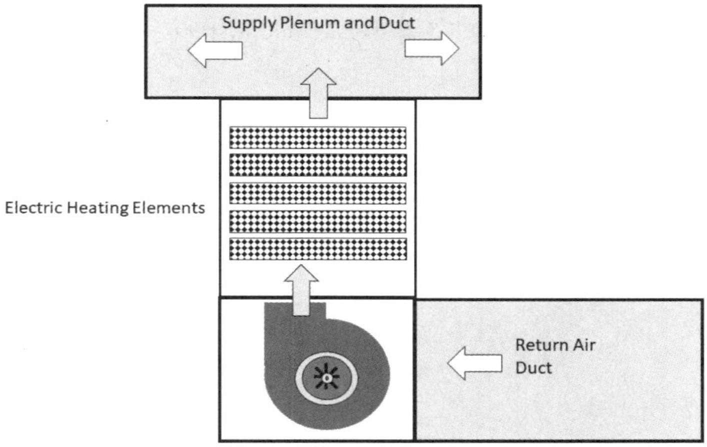

FIGURE 5-6 Electric Furnace

Another electrically energized noncombustion, forced hot air heating system is the heat pump system (Figure 5-7). The heat pump system uses the mechanical refrigeration compression cycle to provide heating or cooling to an air handler's internal Freon coil (evaporator) to accomplish heating or cooling, depending on the season's requirements.

The heat pump system consists of an internal air handler outfitted with an internal evaporator refrigeration coil, a reversing valve allowing the system to change between heating and cooling, and an externally installed condenser unit that houses the refrigeration compressor and condenser coils. The refrigerant (typically R410a) passes through the system, and as it does, it absorbs and releases heat.

Heat pumps are also outfitted with backup electric resistance heaters, much in the same way as an electric furnace is, to provide heat when colder wintertime temperatures exceed the heating ability of the refrigeration cycle.

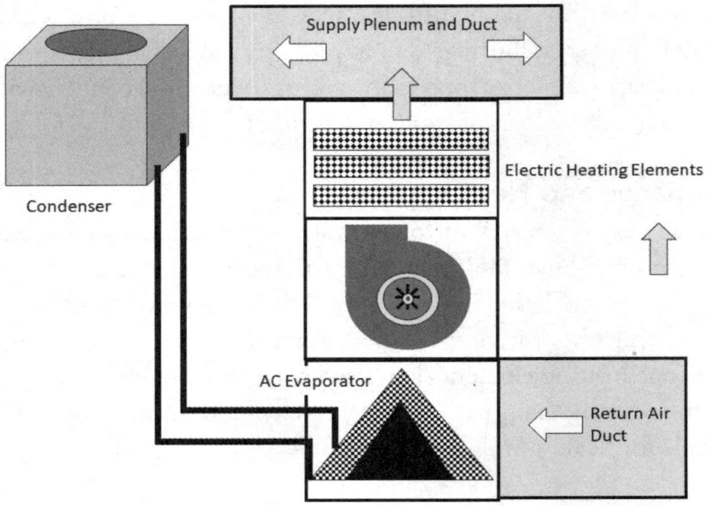

FIGURE 5-7 Heat Pump System

Hydro Air

The hydro air system consists of an internal air handler that is outfitted with a refrigeration evaporator coil and condenser, allowing air conditioning during summer months. In the colder months, heat is produced by a hydronic hot water (forced hot water) coil that receives heat from a centrally installed forced hot water boiler (Figure 5-8).

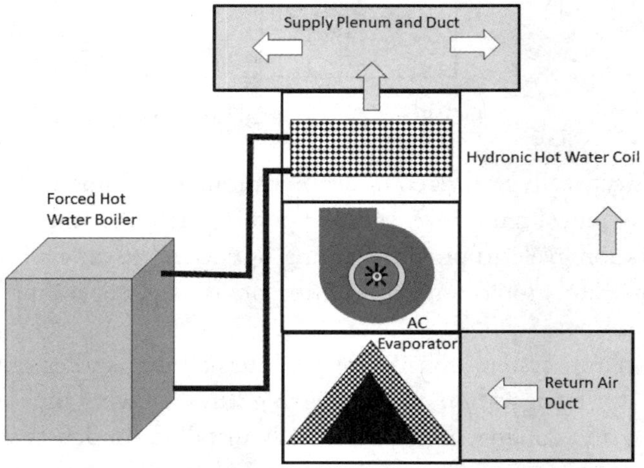

FIGURE 5-8 Hydro Air System Consisting of a Boiler, Hydronic Water Heating Coil and Air Handler

Air handlers are positioned in various areas of the home, and hydronic water and refrigeration line sets connect the refrigeration and hydronic water coils to the condenser and boiler systems (Figure 5-9).

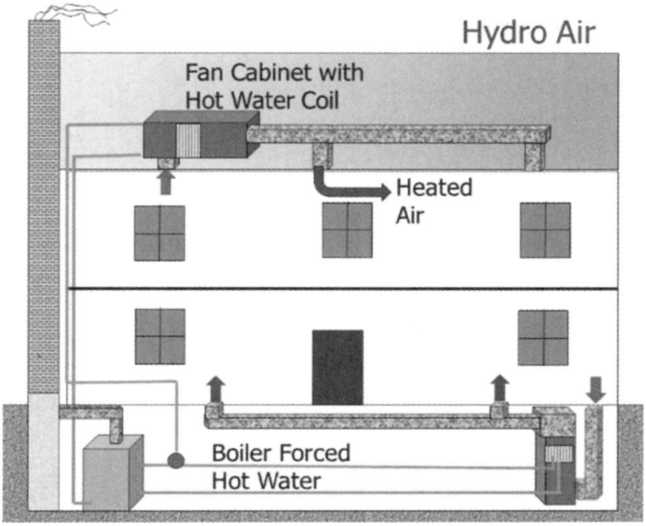

FIGURE 5-9 Hydro Air Components Positioned Throughout House

This system provides the efficiencies of a combustion heat source (natural gas, propane, or heating oil) while offering the flexibility of a single system that can provide both heating and cooling of building air. In addition, because the combustion process is removed from the proximity of the air handling component (unlike in a fuel-fired furnace), soot and other by-products are less likely to be entrained in the system air in the event of a burner malfunction.

Combustion Furnaces

Forced hot air heating systems are typically supplied by furnaces that exchange heat from a product of combustion of a fuel such as natural gas, propane, or heating oil on the combustion side (or primary side) of the *heat exchanger* and the building conditioned air (known as the secondary air) side of the heat exchanger.

Forced hot air systems have the flexibility of providing air-conditioning service as well since the same furnace that is used to burn fuel and generate heat during winter months can also be modified to act as an air-conditioning air handler during summer months by the installation of a fan relay control and air-conditioning evaporator coil in its discharge ductwork.

Counter-Flow and Parallel-Flow Heat Transfer

Aside from the actual construction details of furnace heat exchanger design, an understanding of the factors that increase a furnace's heat transfer thermal ability is important for the service technician to know.

Greater heat transfer efficiencies are realized due to a more uniform temperature difference between the heating (primary side) medium and the secondary (heated) conditioned air. As such, there is better heat exchanger performance with counter-flow heat exchangers compared with parallel-flow heat exchangers (Figure 5-10).

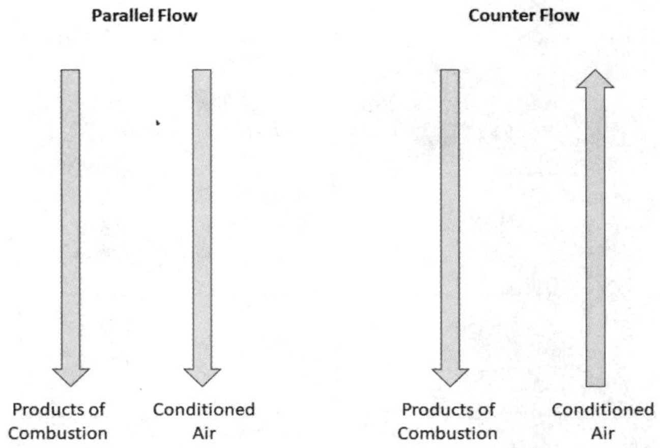

FIGURE 5-10 Counter-Flow and Parallel-Flow Heat Exchanger Directions

Heat Exchanger Flow Path Length

Another factor affecting the ability of a furnace's heat exchanger to transfer heat from the primary side to the secondary side is the surface area between the two mediums.

The surface area of the heat exchanger can be controlled by the length of the combustion flow and conditioned-air paths through the heat exchanger. Greater efficiencies in furnace heat exchangers, which is another way of stating the furnace's ability to use as much of the heat from the products of combustion as possible, can be gained by the design of the furnace's heat exchanger, mainly the product of flow direction and heat exchanger length and surface area.

Heat exchanger design must take into account the factors directing fuel efficiency as well as the effects of the different fuel efficiencies. Higher-efficiency furnaces can absorb so much of the heat from the products of combustion that condensation of these emissions occurs, resulting in acidic liquid accumulation and corrosion.

Furnace Cabinet–Heat Exchanger Orientation

As seen in Figure 5-11, the heat exchanger is located inside the furnace cabinet. This heat exchanger receives the hot products of combustion from the gas or oil burner and discharges them into the exhaust point of the system, such as the chimney inlet or direct vent.

The products of combustion flow inside the heat exchanger, while the secondary or heated (conditioned) air flows inside the furnace cabinet but along the outside of the heat exchanger. In this arrangement, the only interaction between the heating (primary) and heated (secondary) mediums is through heat transfer due to the heated surface of the heat exchanger. The furnace cabinet is insulated, preventing heat loss from processed air (Figure 5-12).

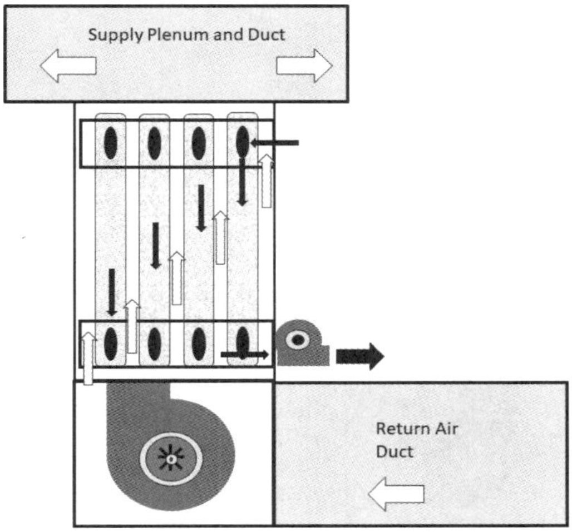

FIGURE 5-11 Heat Exchanger Placement Within Furnace Cabinet

FIGURE 5-12 Furnace Cabinet Exposed Showing Internal Heat Exchanger and Insulated Cabinet

Oil Furnace Design

As seen in Figure 5-13, in the oil-fired furnace, the oil burner is located in the combustion chamber of the heat exchanger. The flue, on the opposite side of the heat exchanger, is connected to the chimney or power venting fan. In the upflow furnace shown in Figure 5-14, the blower fan is visible at the bottom of the photograph.

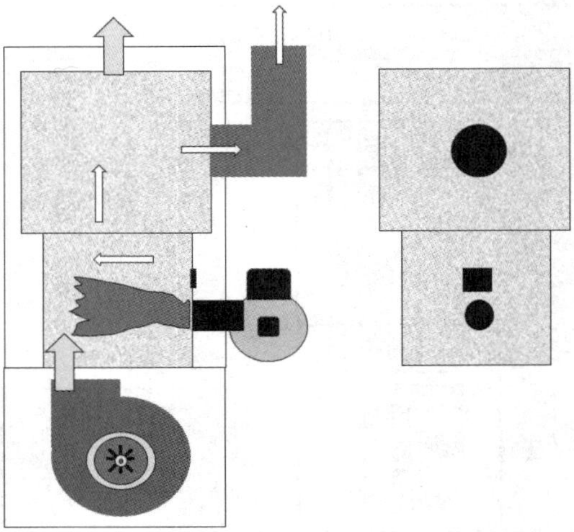

FIGURE 5-13 Illustration of Oil-Fired Forced Hot Air Furnace

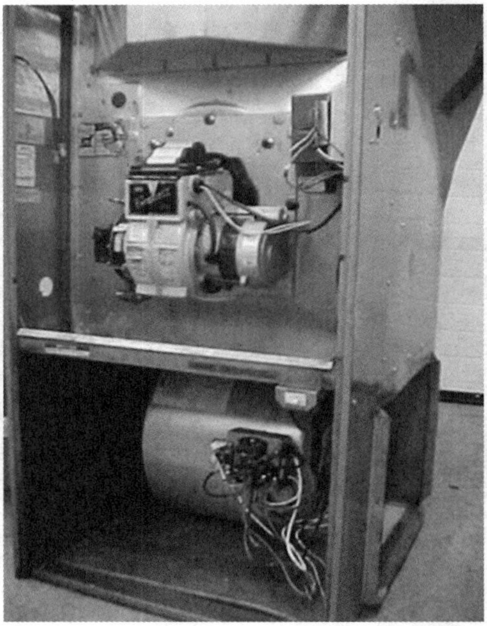

FIGURE 5-14 Oil-Fired Forced Hot Air Furnace. Blower Motor Seen in Lower Cabinet Area

In the oil-fired furnace, the oil burner fires usually into a ceramic fiber–lined combustion chamber. The purpose of the ceramic fiber refractory is to protect the heat exchanger from the direct heat of the burner flame. Without it, flame temperatures are capable, over time, of melting the steel of the heat exchanger and inducing heat exchanger flame impingement holes.

Oil furnace heat exchangers are usually built of welded construction mild steel. In this type of heat exchanger, products of combustion initially flow into the first drum segment of the heat exchanger and then onto the secondary outer heat exchanger through the internal flue passageways.

Figure 5-15 shows a drum-type heat exchanger removed from the insulated furnace cabinet of an oil-fired furnace, where the oil burner inlet is at the base and the discharge connection is at the top. Figure 5-16 shows a bisected drum-type heat exchanger.

FIGURE 5-15 Drum Style Heat Exchanger

Oil-fired furnaces are typically in the 80–84% AFUE range due to the inability to remove additional heat from the products of combustion, which would result in unacceptable condensation issues.

Annual or preventative maintenance of oil-fired furnaces relies on a thorough inspection and cleaning of the flue passageways in the heat exchanger. The heat exchanger is typically outfitted with an observation port above the combustion chamber to allow the technician to inspect the chamber and view the burner flame. Additionally, clean out openings are positioned to facilitate vacuuming and cleaning the heat exchanger flue passageway (Figure 5-17).

FIGURE 5-16 Bisected Oil-Fired Heat Exchanger. Note Positioning of Oil Burner and Combustion Chamber Refractory

FIGURE 5-17 Furnace Cleanout Opening

Gas Furnace Design

Gas-fired (natural or propane) furnaces are designed and built in a different manner from oil-fired furnaces. Design configurations include:

- Parallel flow–counter flow
- Upflow-downflow
- Heat exchanger type
- Efficiency range

Lower-efficiency furnaces in the (80% AFUE range) are outfitted with a single primary heat exchanger, while higher-efficiency heat exchangers are outfitted with a secondary, condensing heat exchanger.

Heat Exchanger Configuration

The different designs of gas-fired primary heat exchangers vary among manufacturers and models. There are presently three basic configurations of primary gas furnace heat exchangers in use with modern furnaces.

Clamshell Heat Exchangers The "clamshell" heat exchanger is a "one-through" heat exchanger in regard to the product-of-combustion flow path through it. This type of heat exchanger uses a stamped sheet metal design that consists of a number of "cells" (Figure 5-18). Each cell is equipped with a gas burner tube or burner, allowing products of combustion to flow through each cell in the heat exchanger assembly. The combustion flow path extends across the entire inside surface of the clamshell heat exchanger, with products of combustion flowing into and filling each cell element (Figure 5-19).

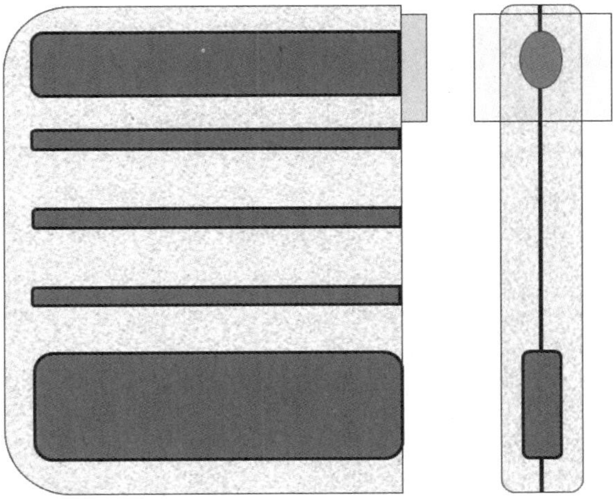

FIGURE 5-18 Clamshell Heat Exchanger

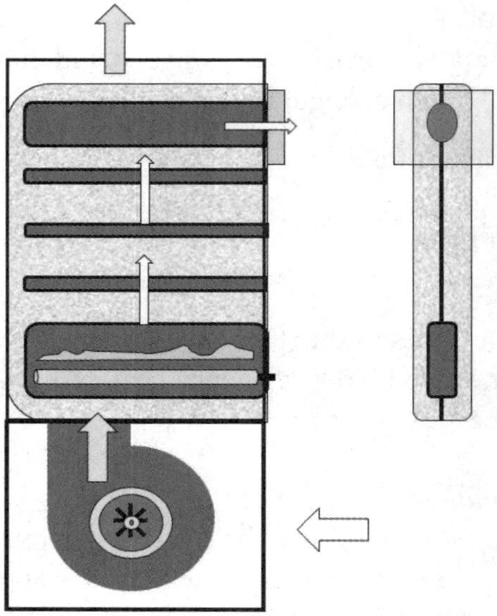

FIGURE 5-19 Furnace Illustration with Clamshell Heat Exchanger

As shown in the photograph of an exposed heat exchanger in Figure 5-20, the heat exchanger battery of cell elements is positioned inside the furnace cabinet. System (secondary) air flows over the heat exchanger cells to and then returning from the various rooms through a system of supply and return ductwork.

FIGURE 5-20 Clamshell Heat Exchanger Exposed

Serpentine Heat Exchangers A major advantage of serpentine heat exchanger design is that it increases the heat exchanger flow path through the heat exchanger. In this type of heat exchanger, two opposing yet symmetrical sheet metal pieces are stamped with a corresponding "S" shape that when joined form a serpentine "S" flue passageway, or channel, through the heat exchanger (Figure 5-21).

FIGURE 5-21 Illustration of Serpentine Heat Exchanger

The edge segments of the heat exchanger are secured by a welding or crimping method. Some manufacturers use a crimp ring, or "eyelet," to secure the inner portions of the heat exchanger adjacent to the serpentine channel (Figure 5-22).

FIGURE 5-22 Serpentine Heat Exchanger

Combustion can be of the natural or induced draft method, allowing parallel or counter flow of products of combustion through the heat exchanger, as seen in Figure 5-23.

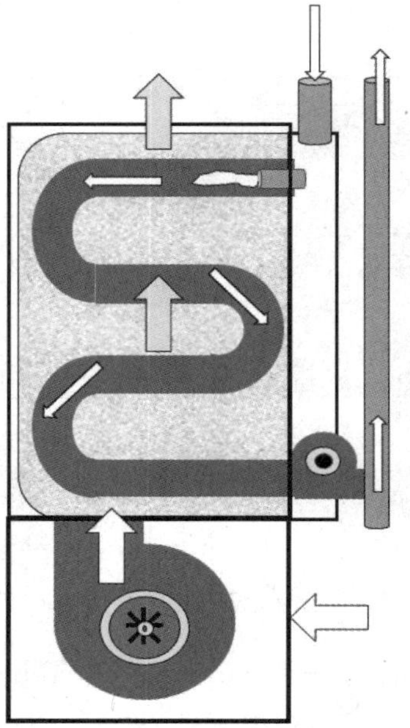

FIGURE 5-23 Furnace Illustration with Serpentine Heat Exchanger

Likewise, as seen in the photograph of an exposed heat exchanger in Figure 5-24, the heat exchanger battery of cell elements is positioned inside the furnace cabinet. This furnace's heat exchanger is discolored due to over-firing of the burners that caused the heat exchanger to overheat slightly in the area where the burners enter it.

FIGURE 5-24 Serpentine Heat Exchanger Exposed

Tubular Heat Exchangers The tubular heat exchanger is a type of serpentine design. It utilizes galvanized steel tubing bent into a serpentine-like "S" pattern through which products of combustion flow, as shown in Figure 5-25. In this type of heat exchanger design, the heat exchanger consists only of the steel tubes, which are arranged in a series of passes and are contained in the furnace cabinet, as seen in Figure 5-26.

FIGURE 5-25 Tubular Heat Exchanger

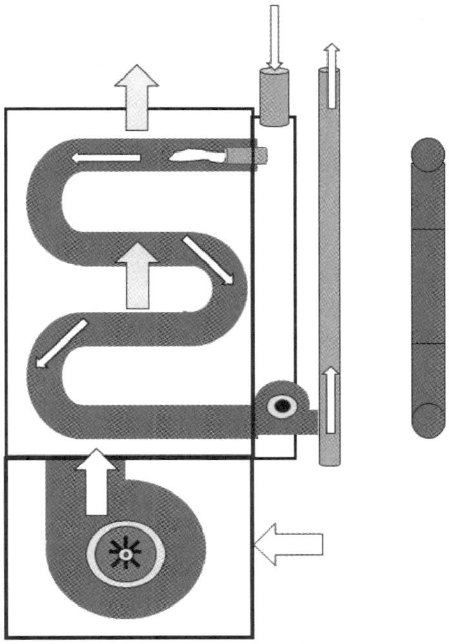

FIGURE 5-26 Furnace Illustration with Tubular Heat Exchanger

The photograph in Figure 5-27 shows the bank of heat exchanger tubes removed from the furnace cabinet. This type of design makes it easier to replace the heat exchanger if needed.

FIGURE 5-27 Tubular Heat Exchanger Exposed

Heat Exchanger Efficiency and Design

As stated, furnaces with only a single "primary" heat exchanger are usually limited to AFUEs in the 80% range. This applies to both gas- and oil-fired appliances. The primary heat exchanger for higher-efficiency furnaces is usually the serpentine or tubular type.

Obtaining 92–98% AFUE by Adding a Secondary Heat Exchanger In addition to the efficiency gained from the counter-flow heat exchanger design, increased efficiency is obtained by increasing the length of the primary heat exchanger. However, even greater efficiencies are obtained by the addition of a secondary heat exchanger, as depicted in Figures 5-28 and 5-29.

As seen in Figure 5-29, the secondary heat exchanger is attached in series to the outlet of the primary heat exchanger, reducing the exhaust temperatures to well below the condensation temperature of the products of combustion. This results in the appliance producing condensate that is a combination of water and condensed combustion gases, which must be dealt with by draining away this corrosive liquid to prevent damage to the furnace.

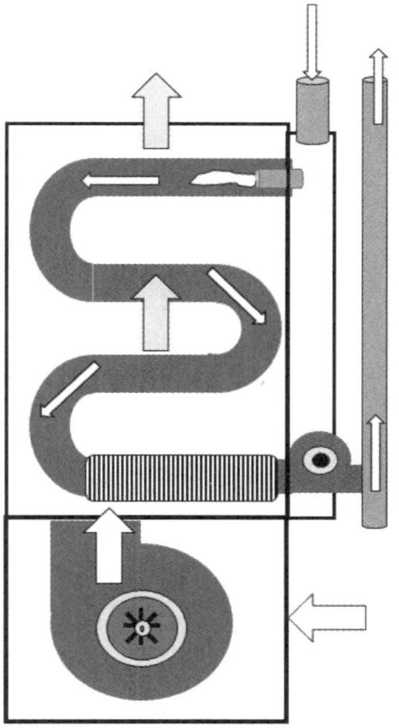

FIGURE 5-28 High-Efficiency Furnace with Tubular Primary Heat Exchanger (Note Secondary Heat Exchanger)

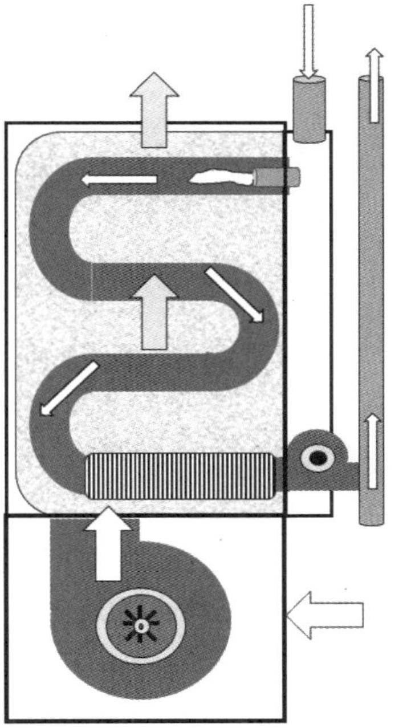

FIGURE 5-29 High-Efficiency Furnace with Serpentine Primary Heat Exchanger (Note Secondary Heat Exchanger)

The secondary heat exchanger, which resembles a copper tube–aluminum-finned radiator, is attached to the outlet of the serpentine primary heat exchanger. See the photographs in Figures 5-30 and 5-31.

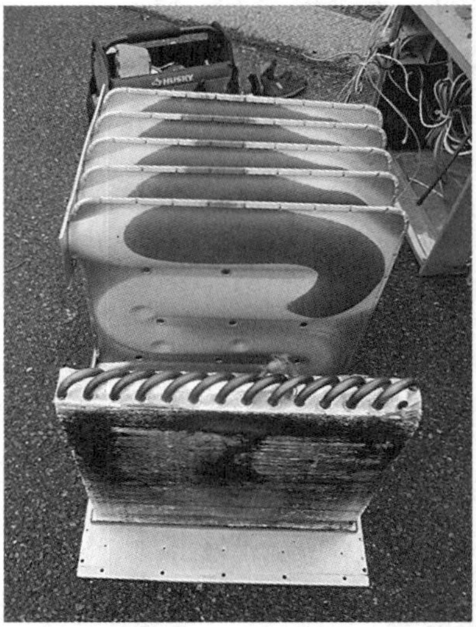

FIGURE 5-30 Disassembled Furnace Showing Serpentine Primary Heat Exchanger and Secondary S-Heat Exchanger

FIGURE 5-31 Secondary Heat Exchanger

Soot Damage Caused by Malfunctioning Furnaces

Is it Really Soot from the Furnace?

Many times, what appears to be soot and is mistakenly thought to be the result of a heat exchanger break may be some other substance not originating from the furnace.

For example, air drawn into dirty ductwork can cause large accumulations of dust and debris to discharge from the ductwork and deposit on the furnace's heat exchanger.

Also, other combustion sources such as fireplaces, candles, and outside air pollutants can be drawn into and distributed through a forced hot air system, giving the appearance of a crack in a heat exchanger. Figure 5-32 shows ductwork that was contaminated with cellulose insulation which insulation contractors accidentally added to the ductwork system.

FIGURE 5-32 Contaminated Ductwork Needing Cleaning

Origins of Soot and Products of Combustion Other Than a Broken Heat Exchanger

There are several other causes of soot and products of combustion besides a broken heat exchanger. As an example, a malfunctioning furnace can add soot to the process airstream; note that this can be addressed by a simple repair.

Other examples:

- Accumulating or depositing products of combustion in the immediate vicinity of the furnace cabinet can be drawn into and through a forced hot air ductwork system.
- An obstructed heat exchanger flue path can cause products of combustion to back up and out of the intended combustion gas path of the appliance.
- Also, as seen in Figure 5-33, furnace cleanout access plates that are improperly sealed have allowed products of combustion to pass through the heat exchanger and into the secondary process air, giving the impression of a cracked heat exchanger.

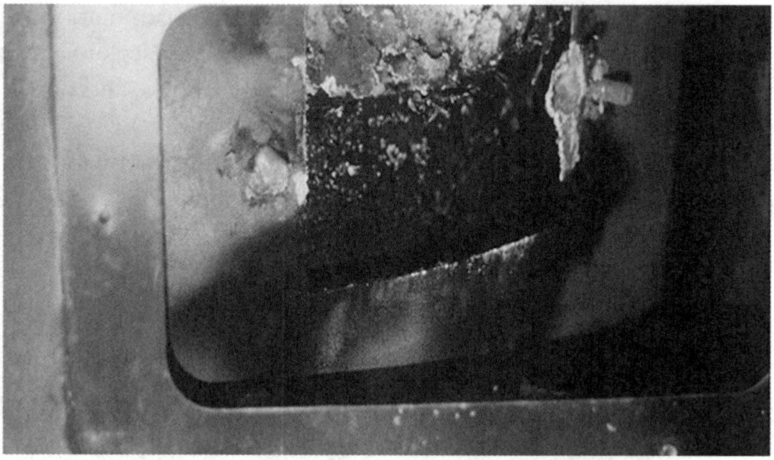

FIGURE 5-33 Leaking Furnace Clean Out Opening

Heat Exchanger Leaks and Cracks

Unlike a crack or other leak in a steam or forced hot water boiler that results in visible water or steam leakage, a break in a forced hot air furnace is more difficult to identify. Heat exchanger leaks (Figure 5-34) can be especially dangerous since products of combustion that can contain carbon monoxide can pass from the primary side to the secondary side of the heat exchanger, mixing with the system air being supplied to the occupied spaces in the home.

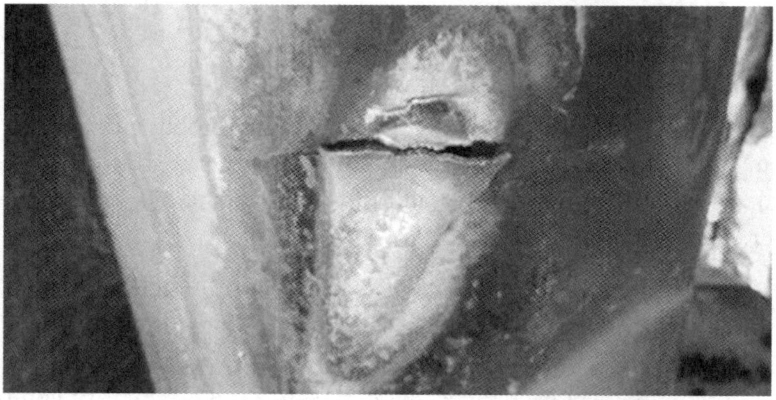

FIGURE 5-34 Overheating Damage to Oil-Fired Furnaces Heat Exchanger

Regardless of the fuel source, all manufacturers recommend at least an annual preventative maintenance check and a simple visual evaluation of the appliance. Issues such as improper firing of the burners or heat exchanger leaks are better caught in a preventative action rather than having to respond to the effects of the malfunction. The philosophy of "If it ain't broke, don't fix it" is the origin of regret and sometimes tragedy in regard to combustion equipment relied upon to maintain heat.

In addition to posing health hazards to the occupants of a home, a leaking furnace heat exchanger can cause soot particles to pass into the structure and damage the interior of the building.

Detecting the Presence of Leaks

The close tolerances between the heat exchanger surfaces, as well as limited access to these surfaces, make finding the *actual* point of leakage in a furnace difficult and sometimes frustrating. The goal of the technician is to find any indication of leakage as well as cracks or signs of impending failure as quickly and efficiently as possible.

Some leaks, such as cracks, holes, or corrosion-induced failures, may be found in obvious and easily accessible locations of the heat exchanger. Others may be hidden from view and only identifiable by inference of their presence through testing.

The first step is to try to identify *if* the furnace *is* leaking products of combustion.

There are several methods of verifying the presence of a break in the furnace's heat exchanger, some more difficult and some more effective than others.

Fan Test

Air movement or communication through the heat exchanger indicative of leaks can be identified by turning on the system blower fan and monitoring the air pressure in the heat exchanger (Figure 5-35). This can be done at the thermostat by switching to the "Fan On" position. Also, some furnaces have their own "Fan On" setting, allowing this to be done directly at the appliance.

FIGURE 5-35 Fan or Blower Test Being Done to Check Furnace Heat Exchanger for Leaks

For oil-fired or sealed combustion furnaces, after the furnace is fired for a few minutes allowing the heat exchanger to be at operational temperature, a draft gauge inserted at the over-the-fire position is used to monitor the fireside air pressure changes, comparing draft gauge readings when the fan is off and when it is on.

For furnaces such as 80% efficient gas-fired furnaces with open ports in the combustion chamber, the use of a draft gauge is not possible. In these instances, the fan test can be done by using a visual indicator of air movement such as talcum powder or commercial test smoke sprays such as those used to test smoke detectors.

Flame Test

Like the fan test, the flame test uses the blower airstream to identify leakage across the sides of the heat exchanger. The flame test can only be used where the burner flame is visible such as in an oil-fired or 80% gas-fired furnace.

For the test, the burner is allowed to fire, and the furnace is observed in its cycles of operation. During the period when the burner flame is firing and the blower fan is on, pay close attention to the flame's shape and movements. Any fluttering or shape change is the result of air movement through the heat exchanger.

The flame test has the advantage over the fan test since it causes the heat exchanger to heat up, potentially opening tighter sharp cracks and enhancing the ability to identify leakage.

Also, signs of excessive flame or heat damage to the burner entry port of a gas-fired heat exchanger (caused by rollout, shown in Figure 5-36) at the observation door to the combustion chamber of an oil-fired furnace may be an indication of a heat exchanger leak. The same signs are also an indication of flue blockage.

FIGURE 5-36 Damage to Gas Heat Exchanger Entry Due to Product-of-Combustion Backup or "Rollout"

Product-of-Combustion Gas Detector

Various combustion analyzers and gas detectors have the ability to detect the presence of products of combustion (Figure 5-37). No matter what the device, care must be taken to verify that it is adjusted with enough sensitivity to identify the presence of these gases since they will be in smaller quantities and diluted in the larger volume of air being delivered through the ductwork.

FIGURE 5-37 Product-of-Combustion Gas Detector Being Used to Verify the Presence of Heat Exchanger Leakage (Courtesy of BOSCH)

The test is done by allowing the furnace to run. The air being circulated through the building by the system blower is monitored with the analyzer for the presence of entrained products of combustion contaminating this air. To check for the presence of leaking combustion gases through the heat exchanger, it is best to sample the air in the ductwork as close to the furnace as possible.

Also, if possible, sample the gases at the ductwork connection to the furnace (plenum). This can be done by cutting a sampling hole through the ductwork, making sure *not* to damage any duct-mounted apparatuses such as an air-conditioning evaporator coil or humidifier. It is also possible in many instances to sample building supply air at supply registers; however, this is best done at registers closest to the furnace in the furthermost "upstream" locations.

Remember, while you can trust the results of a product-of-combustion gas detector test that senses a furnace leak, you cannot assume there is no leak just because the detector does not sense one. Even if the furnace passes a product-of-combustion gas detector test, don't depend on just that result.

Smoke Candle Test

Commercially available "smoke candles" can also be used to detect leakage through a heat exchanger. These candles are essentially scented smoke-generating devices that consist of a fuse and a flammable smoke-generating pyrotechnic-like body.

The test procedure is to bring the furnace up to operating temperature and either to place a smoke candle pointing upward in the furnace's combustion chamber for oil-fired furnaces or to place smoke candles on top of each of the burners for gas-fired upward-firing 80% furnaces. The combustion chamber is then sealed off by closing any observation doors, and the flue connection to the chimney or power venter is blocked off slightly to increase the air pressure in the furnace's combustion chamber.

The furnace can then be fired for a few seconds, causing the positioned smoke candle(s) to light. By observing the heat exchanger through an opening in the plenum, any odor or smoke being discharged can be detected and is an indication of leakage.

Tracer Gas

The introduction of a tracer gas, which is a mixture of compressed nitrogen and 14.3% methane gas, is another method of testing the integrity of the furnace's heat exchanger. This gas can be purchased at heating and air-conditioning or welding supply houses.

This method begins with securing the furnace to ensure that it is not operating. The next step is to make an opening in the furnace plenum ductwork connection on the opposite side of the furnace as far away from the burners as possible. If the plenum is occupied by an air-conditioning evaporator coil, access to the heat exchanger can be obtained either by removing the furnace jacket (if so equipped) or by removing the blower.

A product-of-combustion or gas detector capable of identifying methane is used and should be calibrated to a high setting.

At each cell or chamber of gas-fired furnaces or for the single-combustion chamber of oil-fired furnaces, add the tracer gas, filling the heat exchanger while also positioning the gas detector sampling wand at various locations in the heat exchanger where leakage may be suspected.

MAINTENANCE TIP!

Trusting the Results of Tests

Remember, you can confirm the presence of a leak and condemn a furnace if leakage is detected. However, don't assume that the furnace is necessarily safe if a leak is *not* detected. Factors such as heat exchanger temperature, fan strength, or any crack or opening size and characteristics can mask the fact that heat exchanger is leaking and allow the heat exchanger to pass any of the tests described. Sometimes it is necessary to remove the heat exchanger entirely from the furnace cabinet to visually examine it close up, since many times cracks are no more than a hairline in width and short in length (Figure 5-38).

FIGURE 5-38 Furnace Disassembled Exposing Heat Exchanger

Gaining Access to the Leak (with the Furnace Still Assembled)

Once a leak is identified through the procedures listed above, it is sometimes necessary to gain access to the point of leakage to verify the leak or to satisfy manufacturers' warranties or insurance claims in order to show *why* the failure has occurred. However, the complexity, multiple surfaces, and close clearances between the heat exchanger surfaces make access to the point of leakage difficult.

The technician can use different levels of disassembly of the furnace (from partial to full heat exchanger removal) to facilitate access to the heat exchanger more effectively. It is also important for the technician to make use of past experiences with furnaces to see if there are any common factors in various equipment failures. An awareness of particularities such as make, model, and locations of cracks, corrosion holes, or other supporting damage seen in the past will allow the technician to identify components or areas that will more likely verify the location of suspected failures.

Access to the Heat Exchanger

Gaining access to the heat exchanger to visually inspect the identified point of leakage may require manipulation of the furnace, its auxiliary components, and the ductwork connections.

Plenum Access

Access to the heat exchanger from the top can be obtained by cutting a small access hole in the sheet metal ductwork "box," or plenum, where the heat exchanger and ductwork meet. When the inspection or other work is completed, the access hole can be easily repaired with another piece of sheet metal slightly larger than the opening, held in place with sheet metal screws and sealed with foil tape.

The presence of such an access point is actually beneficial, and in fact, it should be a designed standard of the installation in the first place since it allows easy access to the heat exchanger, increasing the likelihood that a technician will include a heat exchanger visual examination as part of the planned preventative maintenance.

Access to the heat exchanger can be hampered by the presence of an air-conditioning evaporator coil in the ductwork. If this is the case, the evaporator may have to be temporarily removed allowing access to the furnaces heat exchanger. Depending on the equipment layout this may require refrigerant removal from the system. Also, if humidifier housings are installed in the ductwork, they too can hamper access. In this case, access can sometimes be gained through the plenum by removing the humidifier housings.

Blower Removal

Access to the heat exchanger can also be obtained in an assembled and duct-connected furnace through the blower fan opening. This requires the removal of the blower fan, which—depending on the site's installation conditions and appliance design—can vary in levels of difficulty.

Access Through the Combustion Area

Access to one of the most common areas of heat exchanger damage can be gained in the combustion area. The combustion area provides an easy access point to see the appliances combustion chamber refractory as well as a large portion of the heat exchanger. Removal of the oil or gas burner assembly allows an easier and wider view of the portion of the heat exchanger that is exposed to the highest temperatures and harshest conditions.

Removable Jacket

As noted earlier, some manufacturers have designed their furnaces with maintenance in mind by adding a removable jacket section that allows easy and thorough access to and inspection of the heat exchanger. Field modifications of the furnace by cutting and removing sections of the furnace cabinet jacket are not encouraged since internal components such as supports, bracing, and insulation can be affected, posing an operational and safety hazard.

Common Locations of Heat Exchanger Failures

Specific Areas

In standard 80% gas-fired furnaces, a common location for cracks to develop is at the burner entry, behind the rollout shield. Sometimes the crack is not in the heat exchanger itself but rather in a connecting piece of sheet metal acting as the manifold cover of the furnace burner ports in this area (Figure 5-39). Damage in this area will likely be forecasted by scorching or corrosion as a result of heat-induced products of combustion backing out of the furnace.

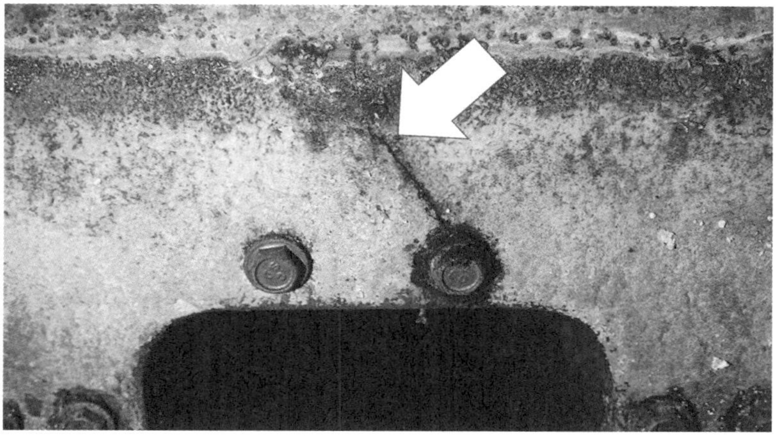

FIGURE 5-39 Thermal Fatigue Crack at Gas Burner Port Entry into Heat Exchanger

The combustion area of the furnace, including the burner entry area in the heat exchanger, is another common location for cracks to develop. This is especially true where the heat exchanger chamber narrows toward the back of the heat exchanger, as seen in Figure 5-40.

FIGURE 5-40 Combustion Area of Furnace's Heat Exchanger

In oil-fired furnaces, the high-temperature insulation (refractory) area in the ceramic fiber combustion chamber must be routinely examined for impending damage since the insulation is the only thing preventing the oil burner flame from eventually impinging directly on the steel of the heat exchanger and inducing failure. The refractory in the combustion chamber is subjected to the erosive effects of the burner flame and hot gas velocity as well as expansion and contraction. This results in the refractory deteriorating over time and becoming paper thin (Figure 5-41).

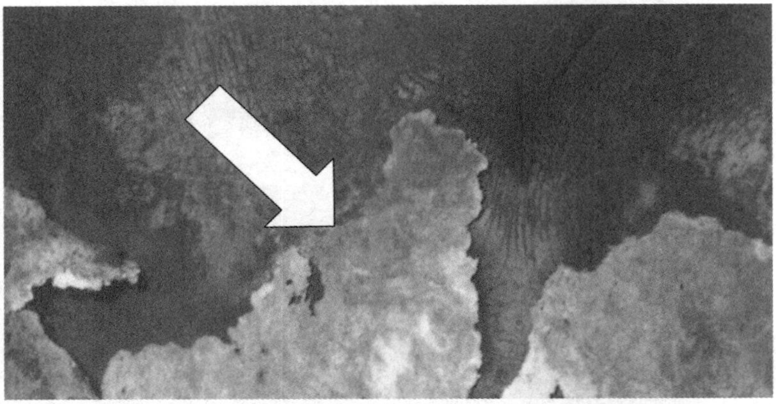

FIGURE 5-41 Thinned Combustion Chamber Refractory Allowed Burner Flame Damage to Heat Exchanger

When the furnace is allowed to operate in this condition for a long enough time, it will allow the burner flame to strike the heat exchanger and cause a burn-through hole to develop (Figure 5-42).

FIGURE 5-42 Flame Impingement Hole in Oil-Fired Furnace's Heat Exchanger Due to Combustion Chamber Refractory Failure

Geometric Shape Changes

Welds in the heat exchanger are subjected to the expansion and contraction corresponding to the heating and cooling experienced during furnace firing cycles. Due to the potential for material inconstancies and brittleness, welds are common locations for cracks (Figure 5-43) that may be difficult to see due to the hairline nature of their development.

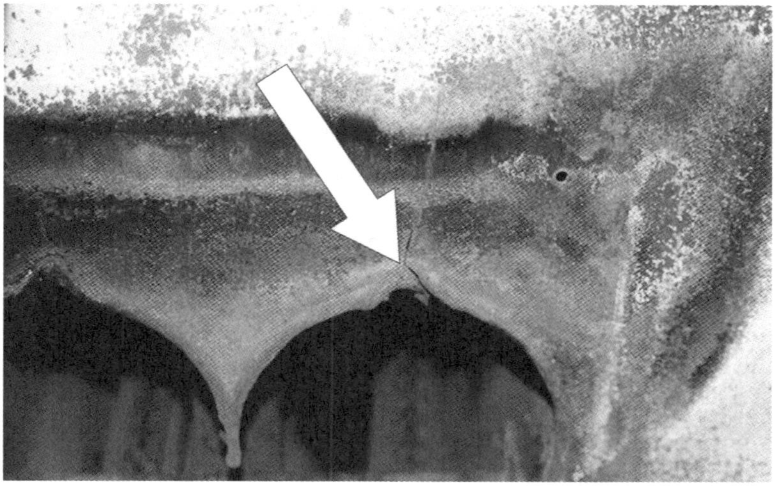

FIGURE 5-43 Crack in Heat Exchanger at Weld Location

Areas such as crimped seams and the approaching bends to these seams (Figure 5-44) are also subjected to expansion and contraction, enhancing the likelihood of cracks developing in these areas of the heat exchanger as well.

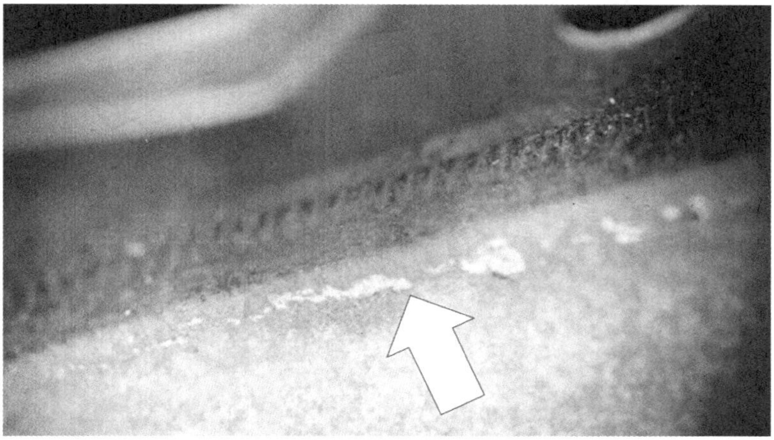

FIGURE 5-44 Crack Development at Bend Area in Heat Exchanger

In serpentine-style heat exchangers (Figure 5-45), the flue passageways are sealed with a "crimp ring," or eyelet, along the flue path, keeping the two pieces of sheet metal together. These fasteners are subjected to the expansion and contraction of the heat exchanger, where the heating and cooling during and after firing cycles can result in damage. Additionally, in a high-heat event, these fasteners will also be subjected to greater thermally induced stress that can cause cracks; as an example, see Figure 5-46, which shows the heat exchanger crack to the left and the separated portion of the crimp ring. Oftentimes, this separation is found in the blower housing of the furnace.

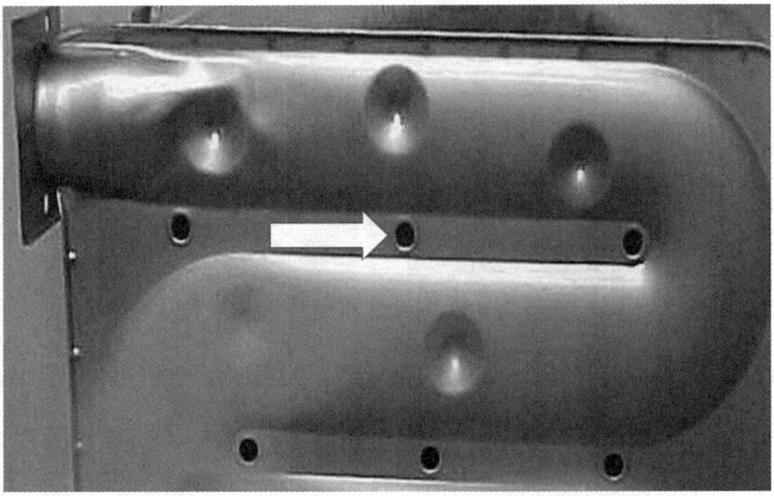

FIGURE 5-45 Location of Crimp Ring in Serpentine Heat Exchanger

FIGURE 5-46 Failed Crimp Ring

On tubular furnace heat exchangers, the tubes themselves are more accessible than stamped sheet metal heat exchangers like the clamshell or serpentine type. Common locations of overheating failures in this type of heat exchanger are also in the portion of the heat exchanger where the burner enters the heat exchanger opening at the approach to the first bend (Figure 5-47).

Another area of focus is the place where each tube is secured to the connecting header, as seen in Figure 5-48. This fastening point, which may be a mechanical or welded connection, can also be checked using a gas detector while the heat exchanger is in operation since this connection point is readily accessible if the furnace plenum is removed.

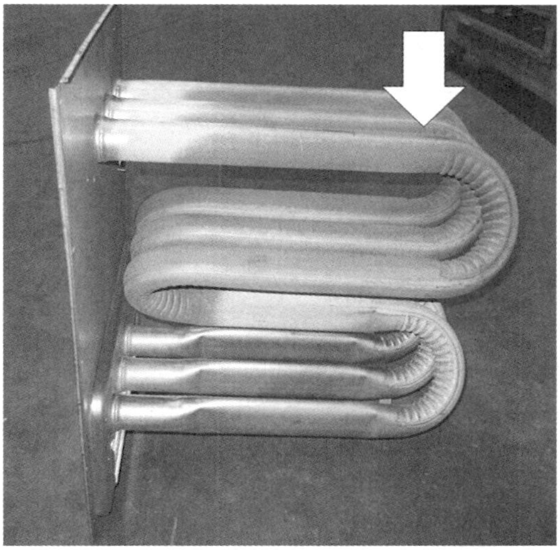

FIGURE 5-47 Vulnerable Area of Tubular Heat Exchanger

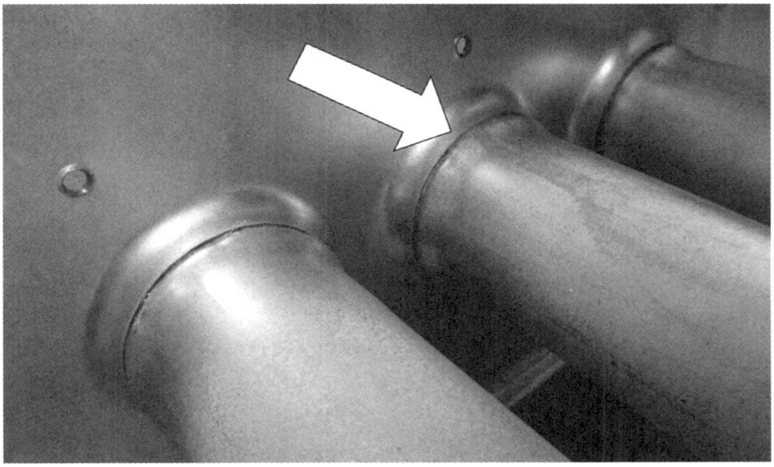

FIGURE 5-48 Connection Point of Tubular Heat Exchanger Tube

Visual and Other Inspection Methods

Now that you know how to detect leaks in the heat exchanger and how different furnaces and their heat exchangers are built, here are some tips and techniques to help you inspect these failures. As you proceed, please keep in mind that not every break in a heat exchanger is a "crack," nor are all cracks caused by the same type of event.

A visual inspection is essential. By gaining access to and actually seeing the failures, the hope is that you can identify *why the equipment failed*, allowing you to make preventative repairs or installation adjustments to keep the same damage from happening again.

Light Test

To do the light test, the furnace should be partially detached from the ductwork to gain access to the air side of the heat exchanger.

The test involves inserting a flashlight into the combustion chamber or firesides (Figure 5-49). The light will illuminate breaks on the opposite (airstream) side of the heat exchanger (Figure 5-50). Often, to enhance the test, the space where the inspection is conducted should be dark, and the light source should be moved to different positions in the heat exchanger.

FIGURE 5-49 Performing Light Test of Heat Exchanger

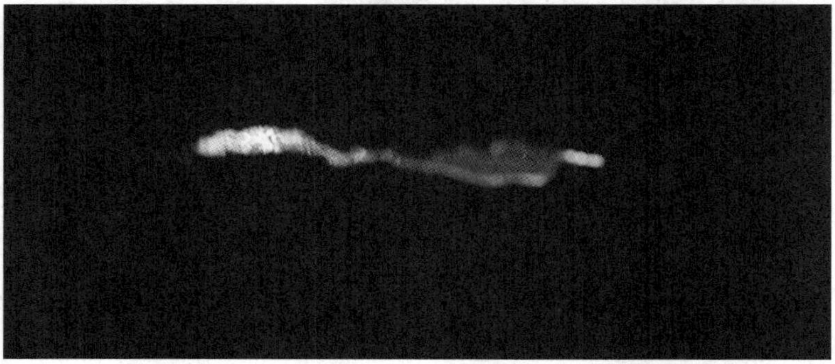

FIGURE 5-50 Illumination Through Heat Exchanger Identifies Point of Leakage

Mirrors

In conjunction with the light test, different sizes of mirrors inserted in the firesides (Figure 5-51) of the heat exchanger can be used to identify heat exchanger breaks. Commercially available inspection mirrors installed on extendable handles may allow access to breaks out of the field of easy vision.

FIGURE 5-51 Inspection of Heat Exchanger with Mirror

Borescopes

Borescopes can be inserted into restricted access portions of a furnace and heat exchanger. These scopes can be used to identify breaks in the heat exchanger. Figure 5-52 is an image from such a device.

FIGURE 5-52 Borescope View

Some borescopes are fitted with cameras. They have the ability to take snapshots or video footage, allowing the point of failure to be shown to the customer or manufacturer pending warranty replacement inquiries.

Oil or Penetrative Dye Test

An oil test is useful for detecting breaks that are not visible even after gaining access to the surfaces. The test is done by spraying (use a spray bottle) a noncorrosive oil on the firesides of the heat exchanger. This results in a dark spot on the air or process side of the heat exchanger indicating where the leak is situated.

This same test can be done with commercial penetrative dyes.

Remember, in spite of the success in identifying a break in the heat exchanger, if there is ever any doubt, *change it out!* Do it even if this entails removal and possible reinstallation or precautionary replacement of the heat exchanger.

FORENSIC INSIGHT

Recognizing Types of Heat Exchanger Failures

Types of Leaks

Heat exchangers leak for many reasons. The spectrum of causes includes everything from crack formation to failed gaskets or seals to holes.

Heat Exchanger Cracks

A crack in a heat exchanger is just that, a crack. It is not a hole of a circular, oblong, or oval shape, and it is usually hairline in nature.

Cracks are not always the result of a singular one-time event. They result from many different causes.

Cracks Caused by Limit Failure, Over Firing, and Overheating

Overheating caused by a high-limit-control failure, ductwork-size imbalance, or over-firing (caused by too large an oil nozzle or gas burner setup) will cause the heat exchanger to take in more heat than it can dissipate. As a result, the structure experiences temperatures greater than it would if it were running normally, causing excessive metal expansion, melting, and other internal stresses.

As an example of a failure caused by overheating, see the photograph in Figure 5-53. The crack is at the end of the combustion path flue of a serpentine heat exchanger. The discoloration and placement of the crack (as indicated by the arrow) are evidence that this furnace overheated. Because of the crack, products of combustion were able to enter into the building's supply ductwork. A failed high-limit control was identified as the cause of the failure.

Cracks Caused by Metal Fatigue

Cracks can also develop in heat exchangers due to metal fatigue. The continual bending forces imparted onto heat exchangers due to thermal expansion and contraction in response to the furnace's firing cycles can cause metal fatigue cracks to develop.

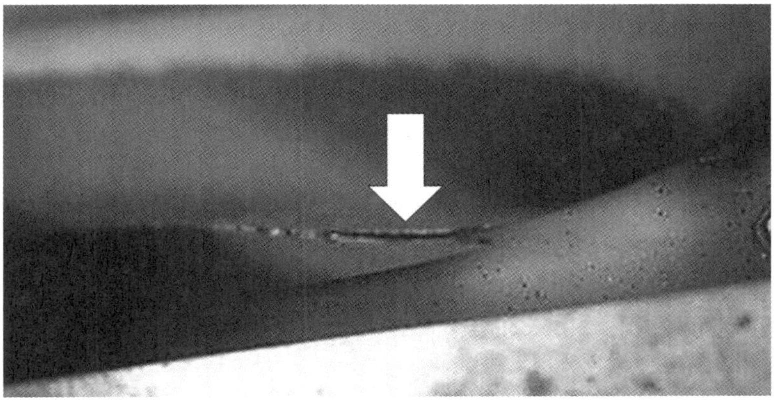

FIGURE 5-53 Overheating Damage to Serpentine Heat Exchanger

There are factors that increase the likelihood of metal fatigue crack propagation, such as the temperature ranges that a portion of the heat exchanger experiences as well as the consistency of the heat exchanger's materials and geometry.

Temperature Range

Like a heat exchanger in a furnace with a failed high-limit control, a furnace installed in a ductwork system that is not balanced may run hotter than it was designed to. Also, if flue obstructions are present, heat exchanger overheating and localized product-of-combustion gas leakage will cause portions of the heat exchanger to overheat. In turn, this causes these areas to experience temperatures in excess of what they were designed for and to expand to levels greater than expected. This increased expansion results in greater bending and other stresses to the heat exchanger in these areas, accelerating the time to achieve fatigue failure of the metal (Figure 5-54).

FIGURE 5-54 Thermal Fatigue Crack in Heat Exchanger

Stress Concentrations

The presence of geometric shape changes, sharp corners, and material inconstancies will also make the heat exchanger vulnerable to metal fatigue cracking since these characteristics create stress concentrations in the structure (Figure 5-55). Stress concentrations are centered at sharp geometric shape changes in the material's design. Elimination of these undesirable features allows uniform (streamline) stress distribution.

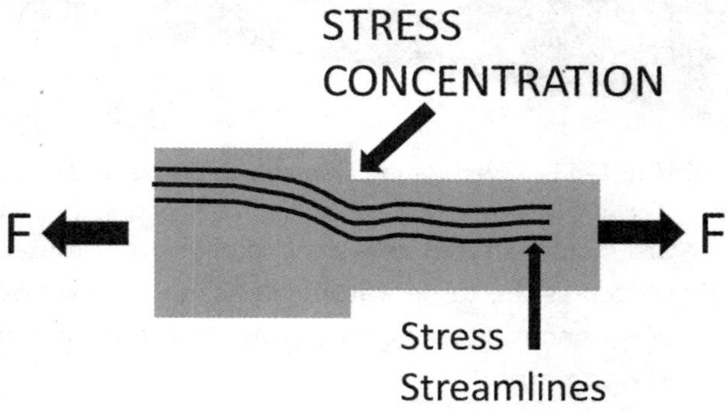

FIGURE 5-55 Stress Concentrations Are Centered at Sharp Geometric Shape Changes in the Material's Design

Conditions such as poor weld joints, slag presence in the weld, or shape changes will be susceptible to metal fatigue failure of the weld joint, resulting in a hairline fracture in the weld (Figure 5-56).

FIGURE 5-56 Heat Exchanger Crack Development in Welded Connection

Rollout Induced

A common location of failure in heat exchangers, due to both an elevated temperature and stress concentration, is at the entry of the burner port to the heat exchanger. In the event that the heat exchanger is not venting properly, product-of-combustion spillage (rollout) causes elevated temperature conditions at the entrance of the heat exchanger. Also, the presence of any weld or design shape inconstancies creates a stress concentration, making the entrance a common location for furnace heat exchanger failure (Figures 5-57). These conditions act together to induce thermal fatigue failure to the heat exchanger.

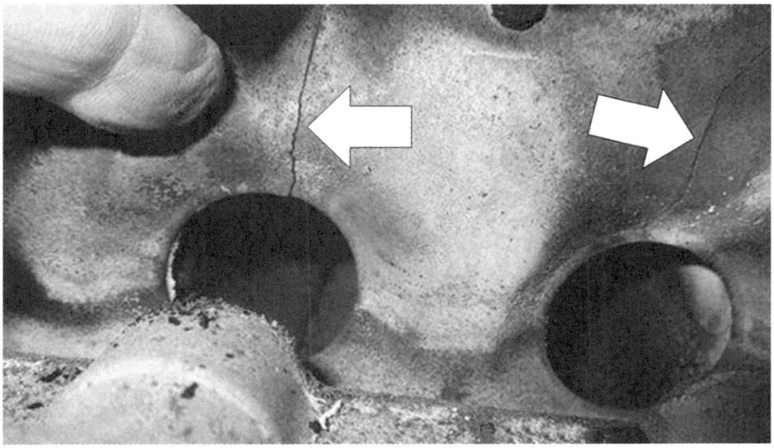

FIGURE 5-57 Thermal Fatigue-Stress Concentration Damage

Flame Impingement Damage

Flame temperatures exceed the melting temperature of steel, and as such, direct contact between the burner flames and the heat exchanger can cause overheating in the area where the flames make contact. Refractory failures, fireside flue blockages, and oversized nozzles are all conditions that can cause the burner flame and combustion gases to be exposed to the heat exchanger, inducing failure.

Flame impingement damage causes melting of the steel, which may begin as rolling or bulges in the steel. Over time, the damaged area will develop into an oval or circular hole with jagged edges (Figure 5-58).

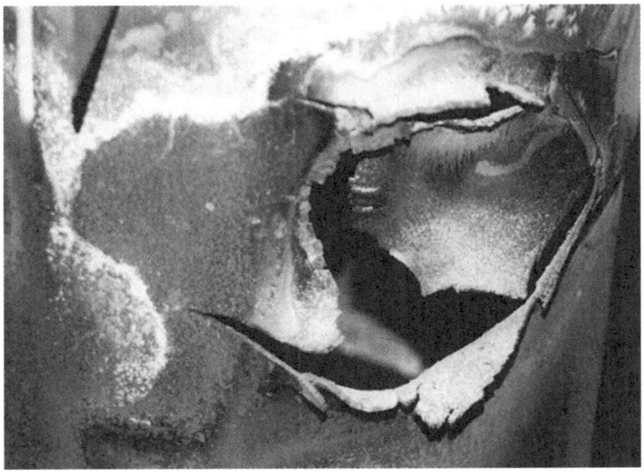

FIGURE 5-58 Flame Impingement Hole in Oil-Fired Furnaces' Heat Exchanger

Corrosion

Water or combustion condensate is corrosive, and if allowed to be in contact with the heat exchanger for a long enough period of time, the liquid will cause corrosion damage (Figure 5-59).

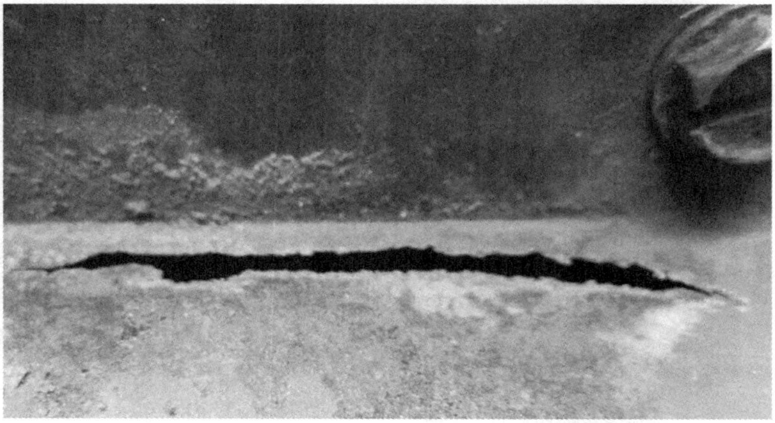

FIGURE 5-59 Corrosion Failure to Heat Exchanger

Common water sources that come into contact with and damage furnace heat exchangers are:

- Leaking air-conditioner evaporator pans
- Malfunctioning duct-installed humidifiers
- Rainwater entering into the exhaust flue

Once corrosion damage appears on a heat exchanger, a progressively degenerative condition begins that will not stabilize itself.

Additionally, when the heat exchanger is damaged by corrosion, there is a greater possibility of stress corrosion cracking (SCC), a cracking process that occurs when a corrosive substance and sustained tensile stresses are present.

TABLE OF FURNACE HEAT EXCHANGER CONFIGURATIONS

Oil fired	Figure 5-13
Clamshell	Figure 5-19
Serpentine	Figure 5-23
Tubular	Figure 5-26
Tubular high-efficiency	Figure 5-28
Serpentine high-efficiency	Figure 5-29

References

Prach, Ellis, *Residential Furnace Inspection Manual,* Heat Exchanger Experts PMB#283, 305 West Magnolia, Fort Collins, CO, 2000.

Vizi, Roger, *Forced Hot Air Furnaces: Troubleshooting and Repair*, McGraw-Hill, New York, 1999.

CHAPTER 6

THE OIL BURNER AND FUEL SYSTEM

The availability and costs of different fuels, as well as the personal preferences of homeowners and installers, direct the choices of which fuel heating systems will be used.

Oil-fired boilers and furnaces require the installation of an oil tank as well as an oil line (or lines) between the tank and burner. This results in a maintenance scope that includes not only the oil burner, boiler, or furnace but the oil tank and lines (Figure 6-1).

FIGURE 6-1 Oil Burner Installation

Oil Burner Design

Oil burners differ among manufacturers, and each burner is configured for the particular boiler or furnace it is connected to. As such, differences in burner components will exist in installations.

The oil burner accomplishes combustion by compressing a liquid supply of No. 2 fuel oil, forcing it through a nozzle where it then is broken into tiny droplets in a process known as *atomization*. This occurs in the oil burner air (aka blast) tube. At this point, the atomized oil is exposed to flowing combustion air while simultaneously being ignited by a spark generated by an ignition transformer. The result is a cone of fire that is delivered to the appliance's combustion chamber (Figure 6-2).

FIGURE 6-2 Oil Burner Flame

Basic Components

Residential oil burner design consists of a motor-driven fuel pump that supplies oil to an oil-atomizing nozzle situated in an air, or blast, tube. The oil is ignited by an igniter spark across electrodes (Figure 6-3).

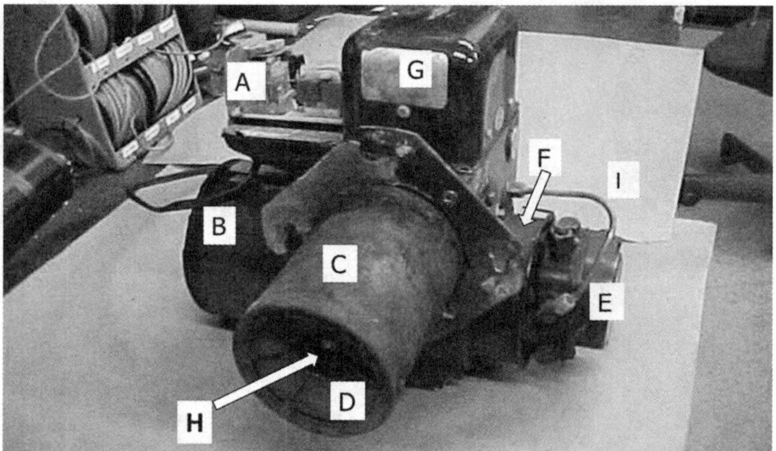

FIGURE 6-3 Major Components of Oil Burner
(A) Primary Control; (B) Motor; (C) Air, or Blast, Tube; (D) Diffuser Head; (E) Fuel Pump; (F) Air Adjustment; (G) Ignition Transformer; (H) Oil Nozzle; (I) High-Pressure "Jet" Line
(Courtesy of R.W. Beckett)

The burner's safety control (A), also known as a primary safety control, prevents the undue accumulation of oil and continual running of a malfunctioning oil burner in the event that the oil spray does not ignite. Each primary control is fitted to accommodate the amount of oil delivered by the oil burner nozzle.

The burner motor (B) is coupled to an internal fan and fuel oil pump (E), which delivers air and fuel through the oil burner's nozzle assembly and blast tube (C). Oil flows from the fuel pump and to the nozzle assembly from a high-pressure oil (jet) line (I) on its path to the nozzle.

The exiting oil flows from a nozzle (H) where it is ignited, and an airflow path is imparted from the exiting gases from a diffuser head (D) that streamlines airflow to enhance complete combustion of the air-fuel mixture.

The Burner Body–Fan Assembly

In the fan assembly (Figure 6-4), combustion air is provided to the fuel atomization mixture from air that is drawn onto the burner through air openings (known as air gates, shutters, or bands) and into the burner housing powered by a squirrel cage fan (A) coupled to both an electric motor and a fuel pump. The connection to the oil pump is made at the connection to the nozzle assembly (C), and the electrical contacts from the ignition transformer are made at the electrode connection points (B).

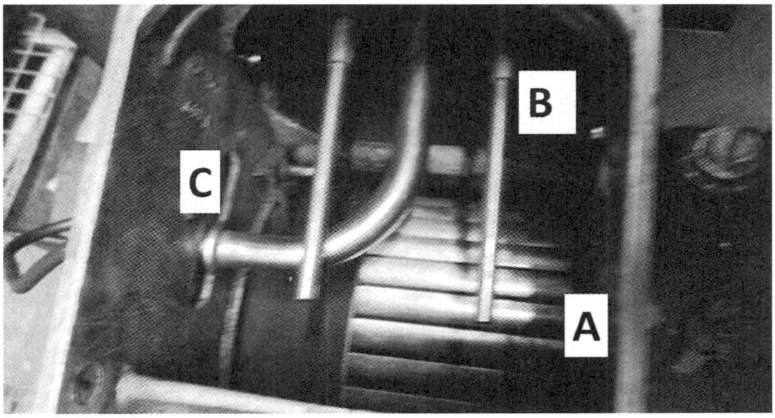

FIGURE 6-4 Internal Components of Oil Burner
(A) Squirrel Cage Fan; (B) Electrode Connection Points; (C) Nozzle Assembly
(Courtesy of R.W. Beckett)

The sketch in Figure 6-5 depicts the orientation of the burner's internal components.

The oil burner body itself is an air-handling vessel, in effect drawing in, compressing, containing, and delivering combustion air to the atomized fuel spray, which also passes through the burner body.

Figure 6-6 depicts the oil burner motor (A) and fan (B) to which the fuel pump is also connected.

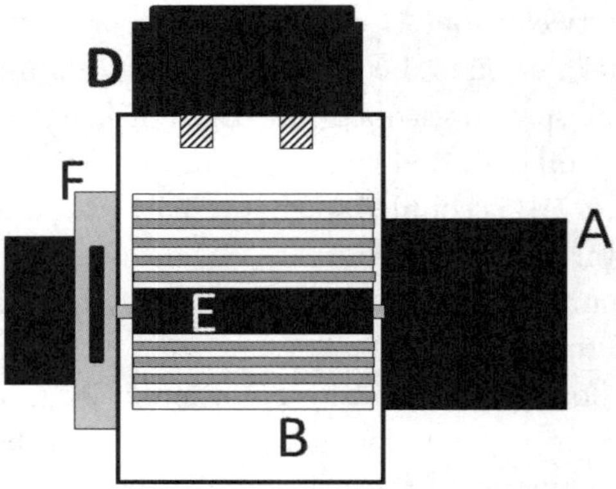

FIGURE 6-5 Oil Burner Internal Components
(A) Motor, (B) Burner Fan, (C) Fuel Pump, (D) Ignition Transformer, and (E) Coupling Between the Burner Motor, Fan, and Fuel Pump, as Well as the Positioning of the Air Entrance "Gate" (F) to the Burner (Courtesy of R.W. Beckett)

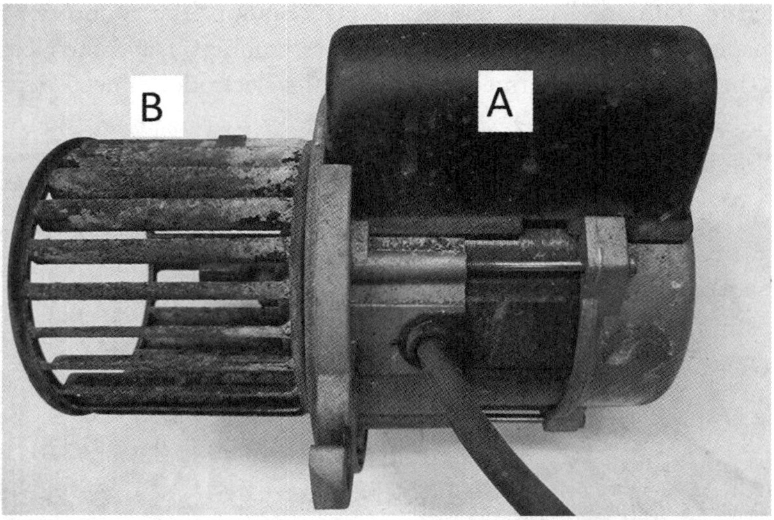

FIGURE 6-6 Oil Burner Motor-Fan Assembly (Courtesy of R.W. Beckett)

As shown in Figure 6-7, fuel oil is provided to the oil burner from the fuel tank and then delivered by an externally mounted fuel pump (A) to a high-pressure oil line (B), where it connects to the nozzle assembly at the burner body.

FIGURE 6-7 (A) Oil Pump and (B) Fuel Line (Courtesy of R.W. Beckett)

The Fuel System

Figure 6-8 shows the nozzle assembly. From the high-pressure oil line connection to the nozzle line (A), pressurized oil flows to the nozzle, where it is simultaneously sprayed and is ignited at the nozzle exit by a spark across the igniter electrodes (B). Both the nozzle assembly oil supply line and the electrodes are placed in the burner blast or air tube (C) that is also cooled by airflow provided by the burner fan.

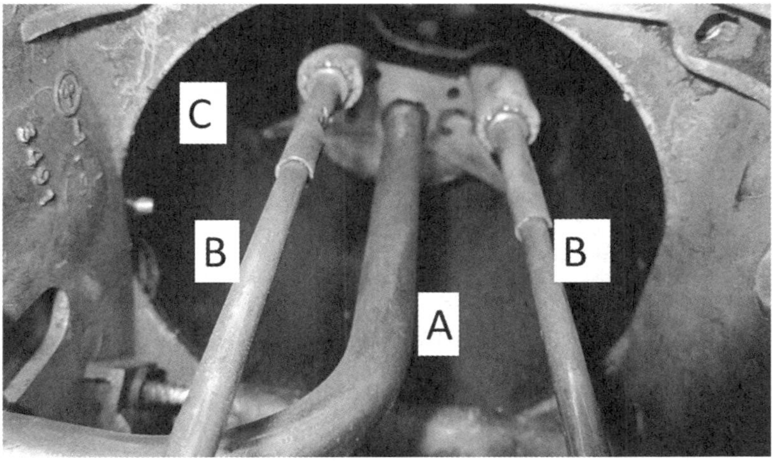

FIGURE 6-8 Components Within the Air or Blast Tube
(A) Nozzle Line, (B) Electrodes, (C) Burner Blast or Air Tube (Courtesy of R.W. Beckett)

As shown in Figures 6-9 and 6-10, the end of the nozzle assembly contains the nozzle (A) and electrodes (B). The gap between the electrodes is a crucial measurement that must be checked during routine maintenance actions.

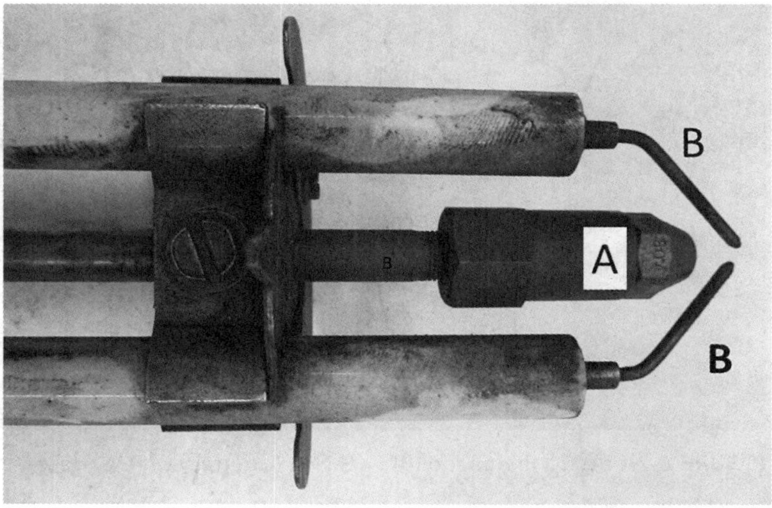

FIGURE 6-9 Nozzle Assembly
(A) Nozzle, (B) Electrodes (Courtesy of R.W. Beckett)

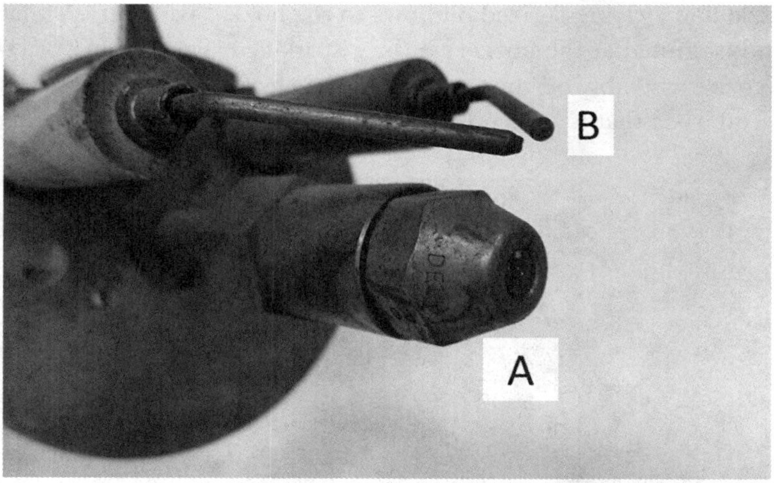

FIGURE 6-10 Nozzle Assembly
(A) Nozzle, (B) Electrode Gap (Courtesy of R.W. Beckett)

The Ignition Transformer and Electrodes

Figure 6-11 shows the ignition transformer and electrodes. The spark that is formed across the electrodes is generated from a 120–10,000-volt ignition transformer (A) positioned on top of the oil burner body. The electrical connections (B) contact the electrodes (C) to supply the voltage necessary to generate the ignition spark.

The strength of the ignition transformer, whether the ferromagnetic or electronic type, can be checked visually as well as by using a commercially available ignition transformer tester, as shown in Figure 6-12.

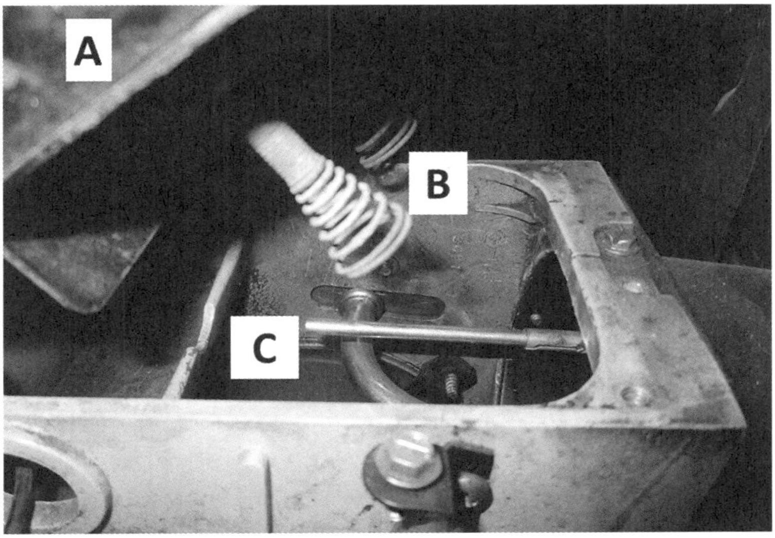

FIGURE 6-11 Ignition Transformer and Electrodes
(A) Ignition Transformer, (B) Transformer Electrical Connections,
(C) Electrode Contacts to Transformer

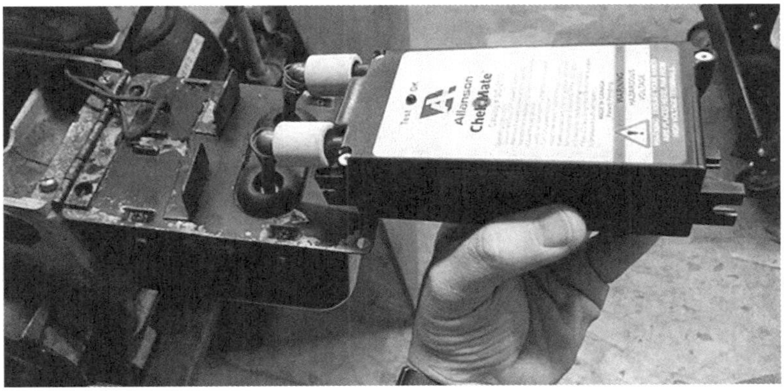

FIGURE 6-12 Ignition Transformer Tester (Courtesy of Allanson International)

Air Path and Air Adjustment

Airflow through the oil burner serves two purposes: It supplies the atomized oil with a combustion air source. It cools the oil nozzle, the primary control CAD cell eye, and the diffuser head, which are exposed to radiant elements of the heat of combustion. See Figure 6-13.

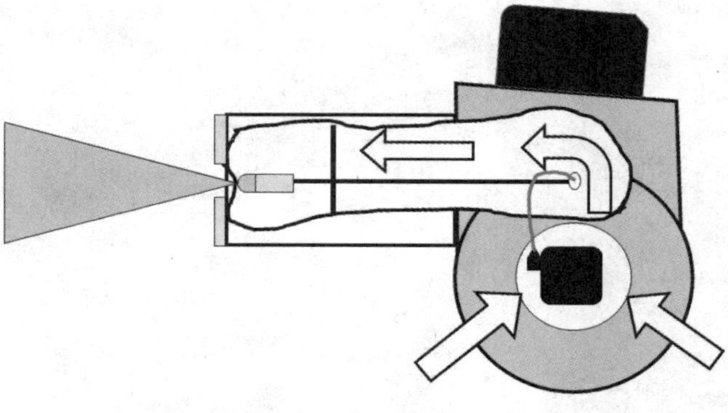

FIGURE 6-13 Air Path Through Oil Burner

Different Manufacturers

Combustion air is adjusted at the air entrance assembly on the outside of the burner. Different manufacturers of oil burners have different burner designs that result in different types of adjustment mechanisms. However, all burners rely on the ability to adjust the air intake to the burner to achieve complete and efficient combustion.

For the Carlin EZ-1 series burner, for example, the air band corresponds to the nozzle size for initial setup, as seen in Figure 6-14.

FIGURE 6-14 Carlin Air Adjustment (Courtesy of Carlin Combustion)

The Beckett burner, shown in Figure 6-15, uses a combination of an air band (B) and air shutter (C) to make larger- and smaller-range air adjustments. The position of the nozzle and nozzle assembly in the burner air tube is adjusted by the Z-dimension adjustment (A), which is also the location where the high-pressure oil line connects to the nozzle assembly.

FIGURE 6-15 R.W. Beckett Air Adjustment
(A) Z-Dimension Adjustment (B) Air Band (C) Air Shutter (Courtesy of R.W. Beckett)

The Riello oil burner (Figure 6-16) features a modular "plug-in" type of primary control and an air adjustment that uses an air shutter (A) that allows for adjustment in combustion air quantities. The plate slides over openings in the burner body (B) at different positions identified by the setting indicator.

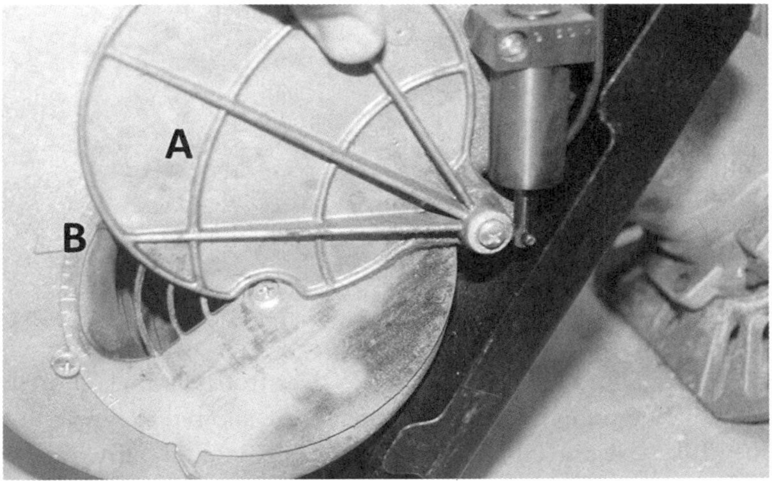

FIGURE 6-16 Riello Air Adjustment
(A) Air Shutter, (B) Air Intake (Courtesy of Riello Corporation)

On any service call or installation, be sure to adhere to the manufacturers' recommendations, and remember that deviations from the manufacturers' recommended practices should only be attempted under their guidance.

Annual and Required Maintenance

All heating appliances require annual preventative maintenance even if this preventative maintenance is a visual inspection of potential problem areas. That said, oil-fired appliances typically require maintenance to a greater level than gas-fired equipment due to the nature of fuel oil combustion.

Filters

Starting at the oil tank, first check the fuel tank filter (Figure 6-17). The filter element requires annual checking and replacement to prevent sludge and other tank impurities from entering the fuel system (Figure 6-18). The filter body may also fail over time so manufacturers' recommendations should be followed to identify the recommended service life as well as compatible components such as filter media and gaskets.

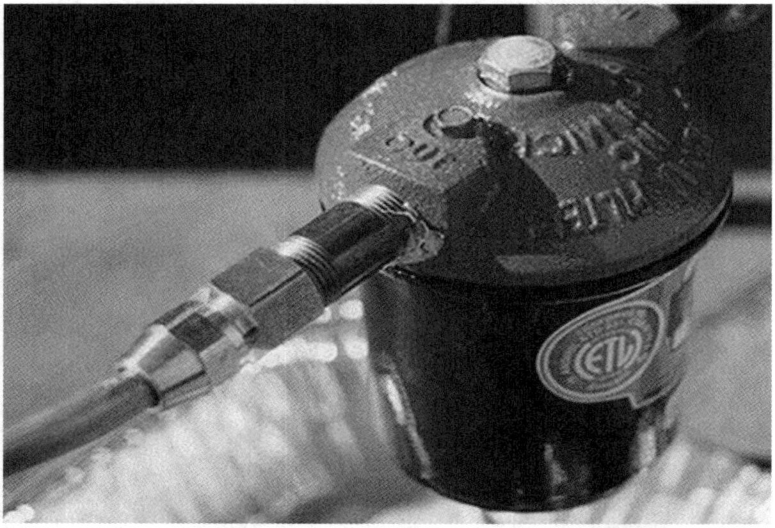

FIGURE 6-17 Oil Tank Filter Housing (Courtesy of General Filters Inc.)

The filter is serviced by closing off the valves from the tank and replacing the filter element, making sure that no contaminants are allowed to flow into the oil line to the burner. Likewise, the oil line between the tank and oil burner can be carefully flushed with oil, collecting the dirty oil in a pail, to make sure any contaminants in the oil line are cleared during annual servicing.

FIGURE 6-18 Moisture, corrosion, and other impurities that enter or are formed in the fuel tank are filtered out and prevented from traveling to the oil burner by the filter's felt or filter medium.

Strainers

In addition to the oil tank filter, the oil burner's fuel pumps are outfitted with a fuel strainer-filter element that is located in the fuel pump housing, as seen in Figure 6-19. Access to the fuel strainer (Figure 6-20) is gained by removing the fuel pump cover.

FIGURE 6-19 Access Cover to Fuel Pump Strainer (Courtesy of Suntec Industries)

FIGURE 6-20 Fuel Pump Strainer (Courtesy of Suntec Industries)

Annual servicing includes replacement of the strainer (Figure 6-21) and fuel pump gasket, reassembly, and a test run to ensure there are no leaks in the fuel pump. Servicing should also include testing the fuel pump pressure and cutoff pressure and doing a vacuum test to ensure that the fuel pump and oil line system are operating correctly.

FIGURE 6-21 Replacing Fuel Pump Strainer (Courtesy of Suntec Industries)

Nozzles

The final component of the fuel system is the nozzle-electrode setup, also known as the nozzle assembly. These components of the oil burner system are the most critical in adjustment and maintenance due to their position at the end of the fuel system as well as their exposure to hot products of combustion.

The nozzle produces a fine spray of tiny oil droplets that form a cloud of atomized oil in the combustion chamber of the boiler or furnace it is connected to. The size of the nozzle (which is determined by the flow rate in gallons per hour), the spray angle, and the cone shape (solid, hollow, or semi-hollow) as well as the fuel pump pressure determine the oil cone and resulting flame geometry. See Figure 6-22.

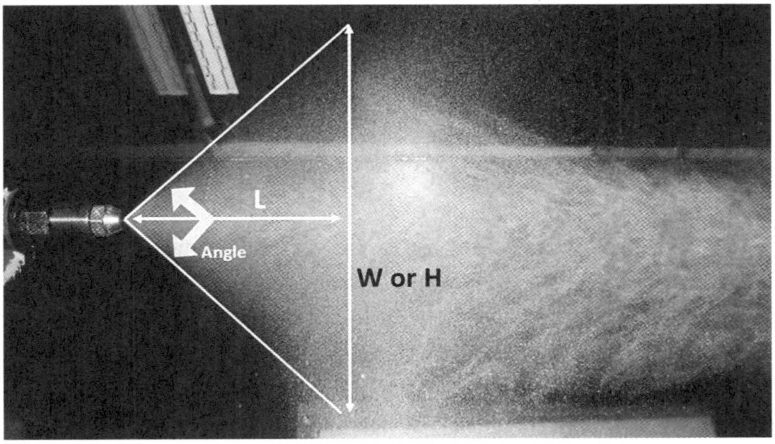

FIGURE 6-22 Oil Burner Nozzle Being Tested for Spray Geometry (Courtesy of ISE Engineering Inc.)

Boiler and furnace manufacturers will typically list nozzle sizes and matching specifications. In addition, nozzle manufacturers will provide performance specifications of their nozzles to assist with nozzle selection. However, for older appliances that have been retrofitted with oil burners, reverse engineering to determine the nozzle size must be performed. This is done by identifying the dimensions and geometry of the appliance's combustion chamber and selecting the appropriate nozzle and pump pressure that will produce a flame of an adequate size. It is important that the burner flame does not exceed the boundary of the combustion chamber, since if it did, this would result in the burner flame striking the chamber walls (known as flame impingement), resulting in soot production and premature appliance flue obstruction.

Clogged Nozzle Diagnosis

If you find that a burner is firing with a soot-laden discharge caused by a clogged nozzle, focus on the cause of the clogged nozzle.

Initial causes of nozzle clogging include:

- No filters in the oil line
- Lack of service
- The oil tank running out of oil, increasing the likelihood of sludge draw
- No sampling hole or combustion test being left behind (and thus no combustion test), allowing a poor firing condition to go undetected

Conditions that can cause poor combustion, which in turn can cause a nozzle to clog, include:

- Boiler or furnace plugged with soot from lack of cleaning

- Fuel pump pressure incorrect
- Air adjustments incorrect
- Electrode setting incorrect
- Weak ignition transformer
- Confined space of installation
- Burner fan fouled (dust, etc.)
- Water leakage into combustion chamber

Electrodes

Both the gap between electrodes (Figure 6-23) and the distance between the nozzle tip and the electrodes (Figure 6-24) must be adjusted in accordance with the manufacturer's instructions to ensure smooth light-off of the burner without any delayed ignition event. Over time, due to the ignition spark erosion of the electrode surfaces, the gaps between the electrodes increase, decreasing the effectiveness of spark generation upon light-off.

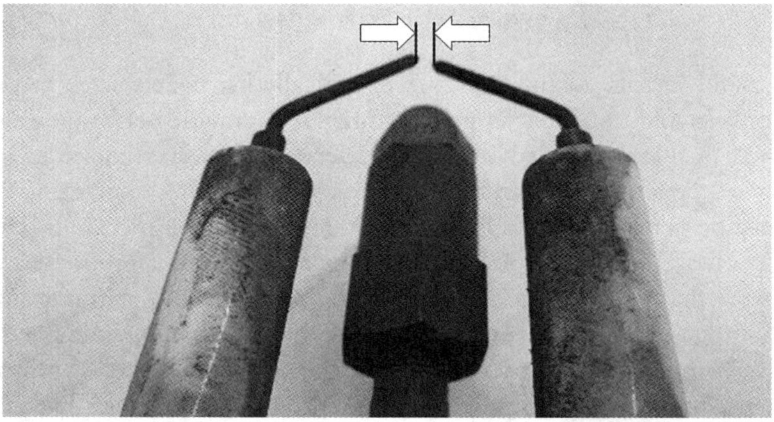

FIGURE 6-23 Gap Between Electrodes

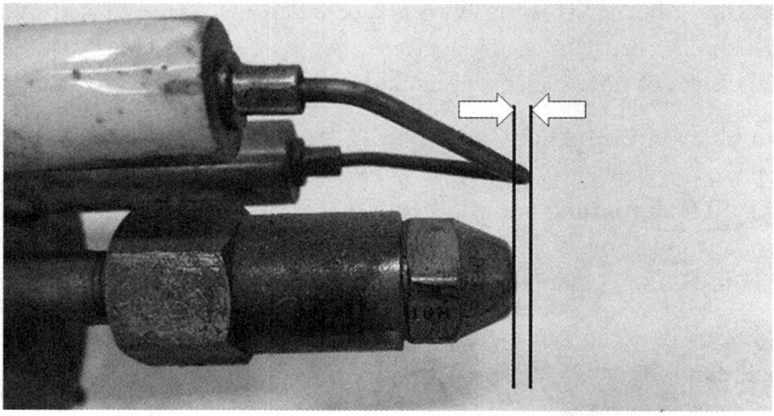

FIGURE 6-24 Electrode Placement Ahead of Nozzle

The Primary Control

The continual flow of unburnt oil into an oil-fired boiler or a furnace's combustion chamber is prevented with the installation of a *primary control* that governs the safe operation of the oil burner. There are essentially two types of primary control technologies in existence today, the older *stack relay* control and more commonly the *CAD-cell* (cadmium sulfide) type of device.

The stack relay primary control (Figure 6-25) is positioned in the connector flue pipe between the boiler and chimney and operates by sensing the presence of heat being discharged from the appliance, indicating that combustion is taking place. Compared with the more modern CAD-cell type of control, the stack relay primary control has a longer time interval (known as a "proving time") between the time the burner initiates a heating cycle and the time it shuts down due to unsuccessful firing. Although the stack relay type of primary control is no longer allowed to be installed in many jurisdictions, many older existing heating appliances still use this type of control.

FIGURE 6-25 Stack Relay Type of Primary Control (Courtesy of Resideo-Honeywell)

The CAD-cell primary control (Figure 6-26) is the basis of all modern combustion control relay devices. It accomplishes flame verification upon the initiation of combustion by the use of a light-sensitive cadmium sulfide photo eye positioned in the oil burner blast tube (Figure 6-27). When the oil burner is energized and combustion is initiated, the CAD-cell eye senses the presence of the burner flame in the combustion chamber, allowing the burner to be continually energized in response to a call for heat. In the event that there is no flame or a poor quality of flame, the residential oil burner primary control will shut down the burner's power circuit after the control's proving time is exceeded.

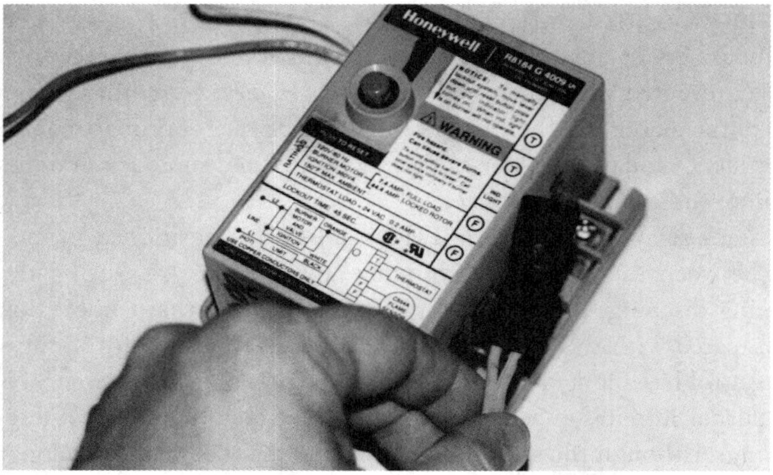

FIGURE 6-26 CAD-Cell Type of Primary Control (Courtesy of Resideo-Honeywell)

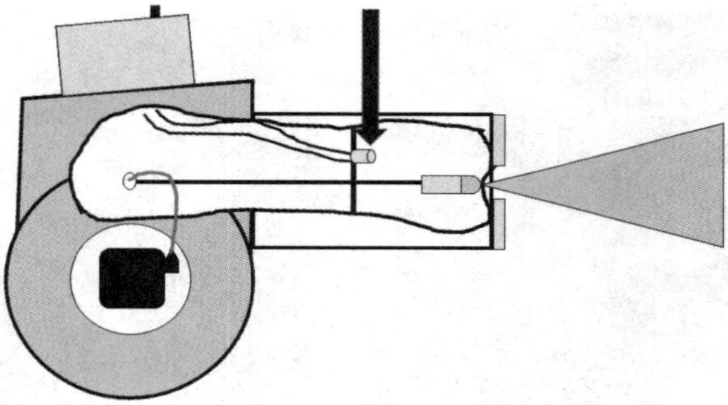

FIGURE 6-27 CAD-Cell Placement Within Oil Burner Air Tube

Various CAD-cell primary controls are available with more advanced capabilities of sequencing burner operating functions. These include the coordinated operation of the ignition transformer, oil control solenoid shutoff valve, and oil burner fan in a manner that allows only air flow prior to the ignition of oil, as well as continued air flow after combustion has ceased. These are done in an effort to provide smooth, soot-free light-off and burner shutdown.

The primary control is typically mounted on the oil burner (Figure 6-28). However, integrated aquastat–primary controls that provide an "all-in-one" control are available.

FIGURE 6-28 CAD-Cell Control Equipped with Pre- and Post-Purge (Courtesy of R.W. Beckett)

The Firesides Cleaning

A boiler or furnace is designed to operate with a clear, unobstructed pathway between the burners and exhaust connector pipe. An obstructed gas combustion path results in harmful product-of-combustion gas leakage as well as equipment overheating and damage. Thus, as needed and as typically recommended by manufacturers annually, the appliance should be serviced,

During annual maintenance, tasks include changing filters, strainers, and nozzles. But by far the most important task is the thorough and proper clearing of accumulated products of combustion (Figure 6-29) in the appliance's flue passageways as well as the combustion chamber and other "fireside" areas.

FIGURE 6-29 Soot-Obstructed Boiler Firesides

Soot removal must be done in a manner that does not cause soot to spread into the home and damage the home's interior. Depending on the design of the boiler or furnace, accumulated by-products of combustion are removed by a combination of vacuuming, brushing, and scraping these deposits from the flue passageways. Soot-removing "swords," or saws, as well as brushes (Figures 6-30) are used to remove accumulated products of combustion from flue passageways and other surfaces exposed to hot combustion gases. A cleaning brush is inserted between sections through passageways between the heat pin field openings, allowing all accumulated by-products to be dislodged (Figure 6-31). Also, specialized vacuums (Figure 6-32) designed specifically for this purpose are used to collect fine particulates.

FIGURE 6-30 Flue Cleaning Brush

FIGURE 6-31 Cleaning Boiler Flue Passageways

FIGURE 6-32 Vacuuming Soot from Boiler's Combustion Chamber

The Combustion Performance Test

This chapter addresses the combustion performance test for oil-fired appliances. However these tests are also required for gas-fired appliances and are further addressed in Chapter 7.

With modern oil-fired as well as gas-fired appliances, gross overall adjustments of a burner can be approximated. However, complete and reliable combustion adjustment requires accurate measurement and adjustment of the combustion parameters.

To analyze how the burner is firing with any accuracy, a combustion performance test must be done. *You cannot do it by eye!*

The combustion performance test (also known as the combustion efficiency test) is actually a set of tests that individually provide diagnostic information about how the appliance is operating but that together yield the final overall combustion efficiency of the appliance. Issues such as the effectiveness of a service call, improper adjustments, and failed burner components can usually be identified and corrected prior to the technician leaving the service call if these tests are performed.

Combustion performance tests must be done not only at the time of the initial installation but also after any service event that may affect how the burner is running. This includes the simple replacement of a nozzle.

Combustion efficiency test equipment has developed significantly in that the older practice of using the chemical fyrite type of analyzer to identify carbon dioxide levels in combustion gases has now been joined by computerized analyzers to identify gas compositions in products of combustion. Likewise, older mechanical draft–airflow measuring manometers or draft gauges can now be replaced with electronic manometers. Regardless of the age or design of the diagnostic equipment, proper calibration and maintenance of these instruments allow them to be reliably used by the service technician.

For oil-fired boilers and furnaces, the following tests are used to evaluate combustion performance.

Smoke

Smoke and soot are the result of unburnt fuel. This unburnt fuel can be caused by issues such as insufficient air for combustion or for oil burners, due to a breakdown in the atomization of oil.

The first adjustment of an oil burner begins with a smoke test to identify if there is enough combustion air. This is accomplished with the use of a smoke pump. The pump, which is operated by hand, flows exhaust gases across a filter paper strip to identify the smoke or soot content of the combustion gases (Figure 6-33). After a set number of full pump strokes (usually 10), the density of the deposit is gauged on a scale of 0–10, with zero to a trace of soot being the target adjustment. See Figure 6-34. Higher soot readings require burner adjustment and attention before the other elements of the combustion performance test can be done.

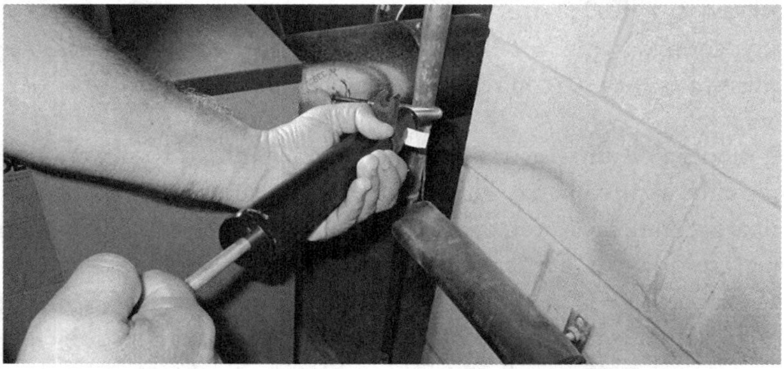

FIGURE 6-33 Performing the Smoke Test

FIGURE 6-34 Smoke Test Results

Draft Measurements
Product-of-combustion gas flow through a boiler's or furnace's heat exchanger, or through the stovepipe connection to a chimney or power venter fan, is measured in inches of water column.

There are two locations for draft measurement. As shown in Figure 6-35, the first is in the combustion chamber of the appliance (B), and the second is in the breach stovepipe connection to the chimney or exhaust fan (A).

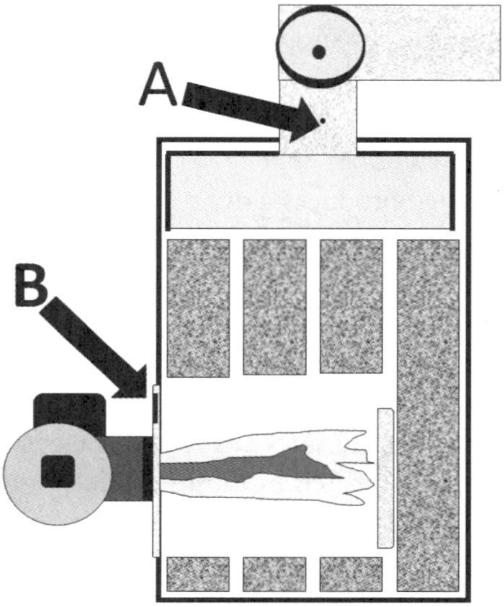

FIGURE 6-35 Sampling Locations to Identify Over the Fire and Breach Draft Measurements

Draft can be measured by a mechanical manometer or draft gauge as well as by an electronic manometer, as seen in Figure 6-36. In this figure, over-the-fire draft is being measured.

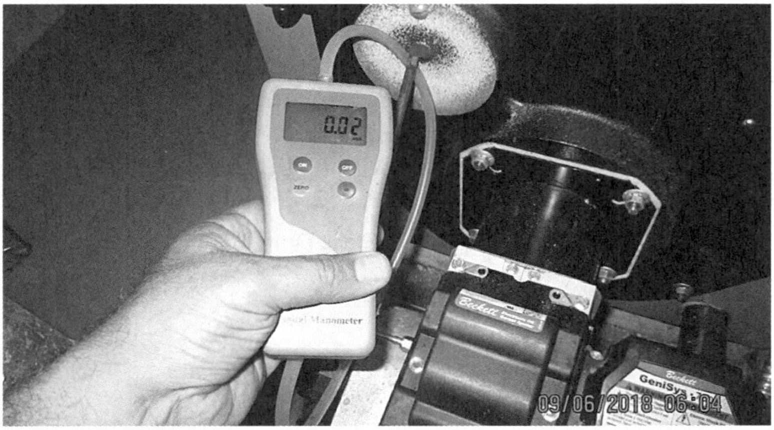

FIGURE 6-36 Electronic Manometer in Use

The resulting measurements are used to diagnose different problems:

- A low over-the-fire draft (below 0.01) accompanied by a low stack draft (below 0.02) is an indication of an improperly adjusted draft regulator, leaks in the flue pipe or chimney, or an improperly sized or blocked flue pipe or chimney.
- A low over-the-fire draft (below 0.01) and high stack draft (over 0.05) is indicative of a clogged heat exchanger.
- A high over-the-fire draft can be the result of a broken regulator, no draft regulator, or an insufficient supply of burner air.

Breach (Stack) Draft

A chimney or combustion chamber that is drawing products of combustion away from the appliance as intended produces a negative draft number. On the other hand, an obstructed chimney or appliance produces no or a positive draft.

A low breach or stack draft (Figure 6-37) can be the result of chimney blockage, a connector pipe that is inserted too far into the chimney flue liner, or an excessively long connector pipe and improper pitch. Breach draft is also affected by chimney height, and as such, shorter chimneys such as those on one-story houses will have a lesser draft than those on a taller chimney will. A typical stack draft is in the range of 0.04 inch of water column, as indicated in Figure 6-38.

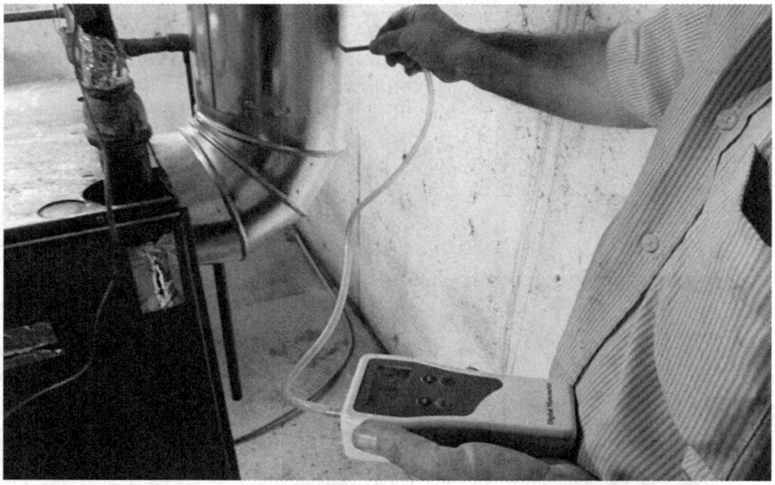

FIGURE 6- 37 Taking Stack Draft Readings

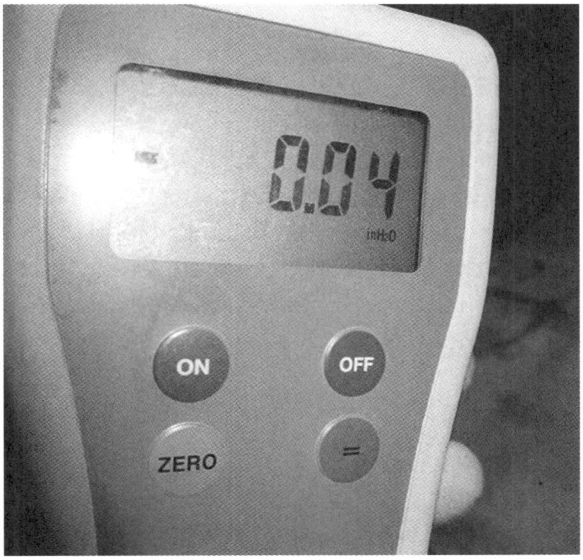

FIGURE 6- 38 Stack Draft Measurement

Stack Temp

Another test that is part of the combustion performance test procedure is the measurement of the breach or stack temperature (Figure 6-39). A high stack temperature above 600°F is indicative of excess draft as well as a soot obstruction in the heat exchanger, preventing the transfer of heat to boiler water or furnace primary air. Another cause of a high stack temperature is the result of too large a burner flame caused by a malfunctioned or overfired nozzle.

FIGURE 6-39 Stack Temperature Measurement

A low stack temperature (below 380°) is an indication of an underfired (too small) nozzle.

CO_2 or O_2

The CO_2 (carbon dioxide) level in the flue gas is an indication of the efficiency of the combustion in the appliance. When the level of CO_2 in a small amount of excess air is high, the heat lost up the flue will be at a minimum.

The level of O_2 (oxygen) or CO_2, can be identified by the Fyrite combustion analyzer (Figure 6-40) or an electronic combustion analyzer. Much in the same way that the smoke test pump is used to extract combustion gases from the boiler stovepipe, the Fyrite type of analyzer uses a hand pump to draw in discharging combustion gases. The fluid will react with the products of combustion and respond by a change in volume to indicate the level of CO_2 or O_2 in the burner emission products.

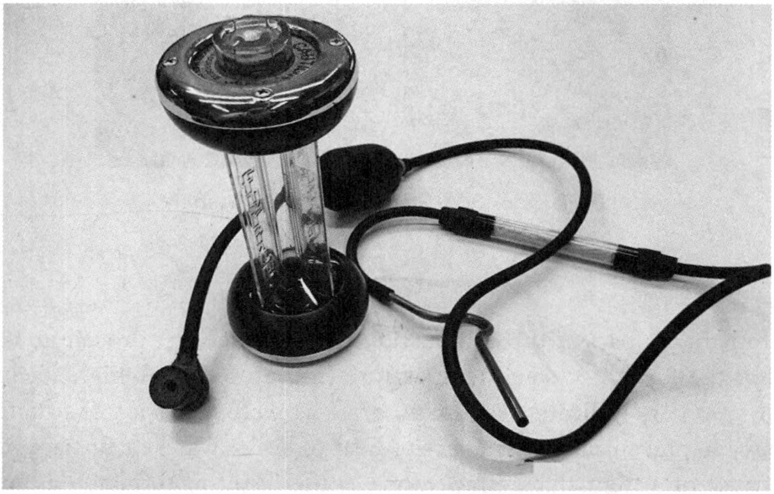

FIGURE 6-40 Fyrite Combustion Analyzer

Computerized Combustion Analysis Equipment

Computerized combustion analyzers are popular because technicians can use just a single instrument to simultaneously perform many of the tests that make up the combustion performance test. See Figure 6-41. In addition, the analyzers are convenient, portable, and accurate. They do require periodic maintenance and factory calibration.

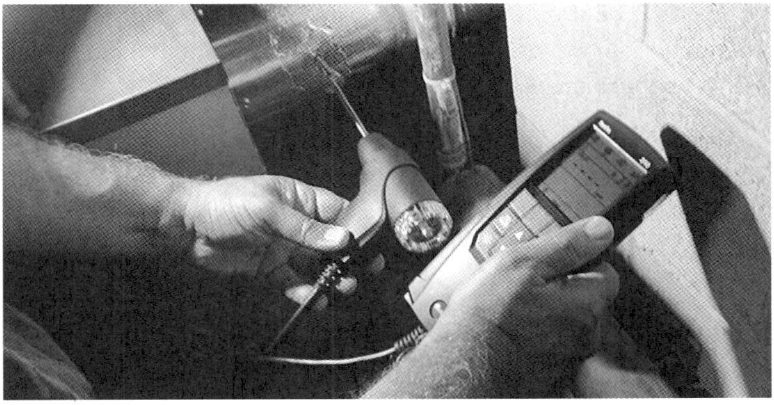

FIGURE 6-41 Combustion Testing Equipment (Courtesy of TESTO Inc.)

Typical Flame Retention Oil-Fired Burner Appliance Readings

Manufacturers will list parameters and combustion target ranges of operation for their appliances; however, following are typical ranges for oil-fired appliances.

- Stack temperature, 60–79% AFUE: 400°–600°F
- Stack temperature, 80% AFUE: 330°–450°F
- Stack temperature, 90% AFUE: Less than 125°F
- O_2: 3–6%
- CO_2: 10–12.5%
- Over-the-fire draft: 0.02-inch water column
- Stack draft: 0.02–0.04-inch water column
- CO: Less than 50 ppm
- Smoke spot test: 0–trace of soot

MAINTENANCE TIP!

More Than Just Efficiency . . . It's Diagnostic

Putting It All Together

The ability to interpret combustion performance test results is in itself something that is gained over time and is based on experience and the information available in specific readings for particular installations of heating equipment. In fact, entire books are devoted to the performance and interpretation of combustion performance testing results.

Also, combustion performance tests are not only done on oil-fired equipment. Elements of these tests yield a final efficiency of the combustion parameters of the appliance for oil, propane, and natural gas as well as wood- and biomass-fired equipment.

Regarding oil-fired equipment, however, certain "target" readings exist that the technician will use to make baseline adjustments. For oil-fired boilers and furnaces these are typically:

- Smoke spot test: Zero to a trace of soot
- Over-the-fire draft: 0.02 inch of water
- Stack draft: 0.02 - 0.04 inch of water
- CO_2: 10–12.5%

Some common symptoms and causes of burner issues may become evident with experience and a familiarity with particular equipment history. These symptoms include:

- **Smoke over No. 1 on new burners or over No. 2 on burners more than 10 years old that also read a high CO_2 level over 12%.** This is an indication of insufficient burner air, which may be caused by an incorrect air adjustment, a dirty fan, or obstructed air intake openings. This can also be caused by a dirty or obstructed air tube and insufficient draft.
- **Low CO_2 under 10%, accompanied by low smoke or a CO_2 reading under 8% with a No. 1 smoke reading.** This is an indication of excess burner air. Aside from the air adjustment, possible sources of excess air include air leaking into the furnace and excess draft.
- **Excess smoke and low CO_2 levels (less than 10%).** This is symptomatic of a faulty or damaged nozzle. Or there may be debris in the combustion chamber or oil spraying onto the end of the air tube or walls of the combustion chamber.

In addressing these problems, always consult the manufacturer's specifications for adjustment.

Service Records

Combustion Performance Test Data

A combustion performance test report slip is proof that you did your job to the best of your ability when you serviced the appliance. It sends the message that the burner was tuned and adjusted with utmost diligence when you serviced it last. It may also save you from liability in the event of an equipment malfunction that results in building damage.

If you truly did perform a combustion performance test, as you should for all installations, as well as perform other ongoing maintenance actions, don't keep it a secret! Print it (for electronic analyzers, Figure 6-42) or write it out and post it on the appliance (Figure 6-43).

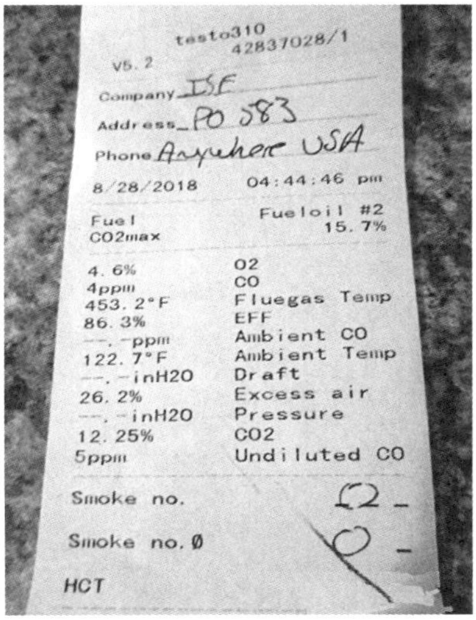
testo310
42837028/1
V5.2
Company ISF
Address PO 583
Phone Anywhere USA
8/28/2018 04:44:46 pm
Fuel Fueloil #2
CO2max 15.7%
4.6% O2
4ppm CO
453.2°F Fluegas Temp
86.3% EFF
--.-ppm Ambient CO
122.7°F Ambient Temp
--.--inH2O Draft
26.2% Excess air
--.--inH2O Pressure
12.25% CO2
5ppm Undiluted CO
Smoke no. 2
Smoke no. Ø 0
HCT

FIGURE 6-42 Electronic Combustion Performance Test Data

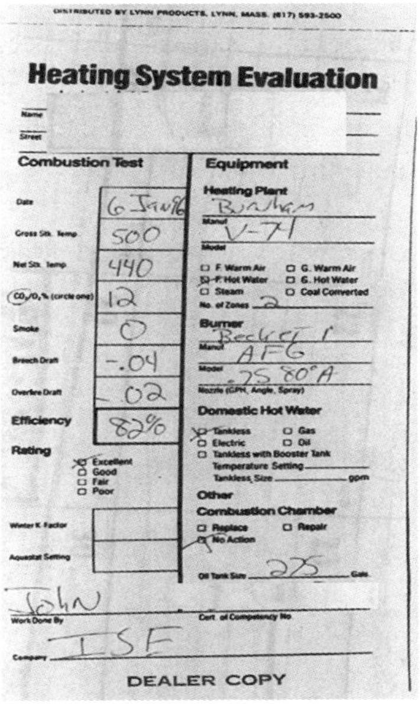
DISTRIBUTED BY LYNN PRODUCTS, LYNN, MASS. (617) 593-2500

Heating System Evaluation

Name
Street

Combustion Test
Date: 6 Jan 96
Gross Stk. Temp: 500
Net Stk. Temp: 440
CO_2/O_2% (circle one): 12
Smoke: 0
Breech Draft: -.04
Overfire Draft: .02
Efficiency: 82%
Rating: ☒ Excellent ☐ Good ☐ Fair ☐ Poor
Winter K Factor
Aquastat Setting

Equipment
Heating Plant
Manuf. Burnham
Model V-71
☐ F. Warm Air ☐ G. Warm Air ☒ F. Hot Water ☐ G. Hot Water ☐ Steam ☐ Coal Converted
No. of Zones 2
Burner
Manuf. Beckett
Model AFG
Nozzle (GPH, Angle, Spray) .75 80° A
Domestic Hot Water
☒ Tankless ☐ Gas ☐ Electric ☐ Oil ☐ Tankless with Booster Tank
Temperature Setting
Tankless Size gpm
Other
Combustion Chamber
☐ Replace ☐ Repair ☒ No Action
Oil Tank Size 275 Gals.

Work Done By John
Cert. of Competency No.
Company ISF

DEALER COPY

FIGURE 6-43 Handwritten Combustion Performance Test Data

Service History

Recording the details of your service actions is important in facilitating continuous reliable operation of equipment. This may be the first time a service company is servicing an appliance, and additionally, service companies have several technicians. In the event of a late-night emergency call when the office is closed or other technicians may not be reachable by phone or text, wouldn't it be useful to have a complete service history posting on the appliance, attached when it was last serviced, as well as a full history of past maintenance? See Figure 6-44.

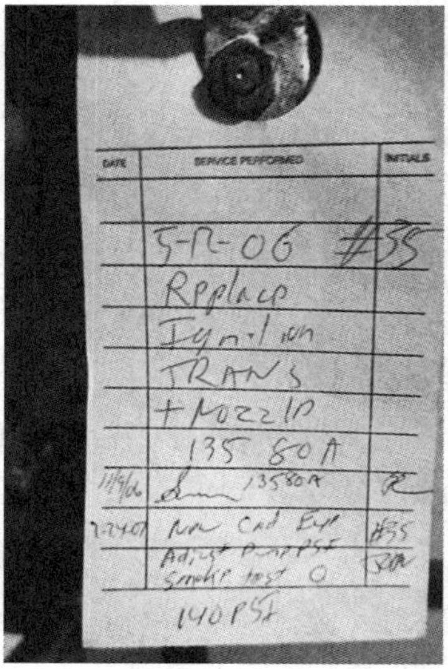

FIGURE 6-44 Details from Last Service Event

Avoid or at least supplement "Cleaned and Checked" preprinted service tags (Figure 6-45) that do not provide any additional detail about what was done during the last service call. These forms may encourage laziness, causing the technician to provide incomplete maintenance information.

FIGURE 6-45 "Cleaned and Checked" Service Tag

Think of it this way. What would *you* like to see if *you* were assigned an emergency no-heat call at 2 a.m. and you hadn't been to the home before? Well-documented service history cards complete with the last combustion performance test (Figure 6-46) provide servicing technicians with a thorough background of the past maintenance, allowing them to better diagnose common equipment-specific problems and to better plan ongoing maintenance actions.

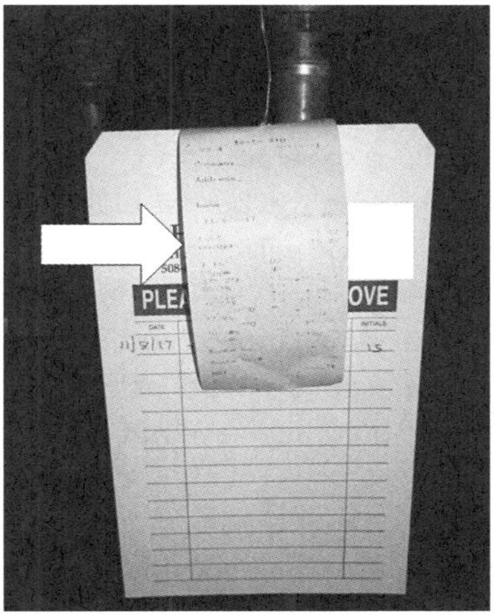

FIGURE 6-46 Complete Service History Listing Work Dates, Work Done, and Combustion Analysis Data

FORENSIC INSIGHT

Overheated Diffuser Heads

The air and product of combustion path from the air inlet of the oil burner, through the appliance and to the connecting breaching into the chimney is a system in itself that the appliance is initially set up to operate with and later maintained to preserve the integrity of this system. Blockage, as depicted in Figure 6-47, disrupts this.

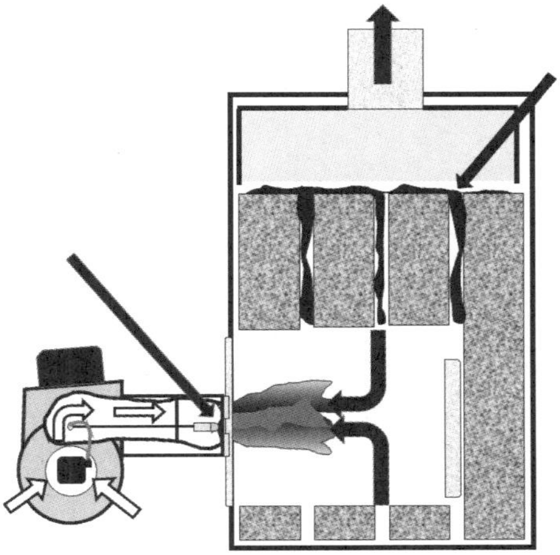

FIGURE 6-47 Obstructed Boiler Flue Passageways

Blockage can be the result of poor or overdue fireside cleaning or a poorly adjusted oil burner that went unnoticed.

Blockage downstream of the oil burner reduces airflow from the burner fan into the appliance. It also reduces the amount of air flowing through the blast tube and over the nozzle assembly. This air is needed not only for complete combustion but also for cooling of the oil burner nozzle. Incomplete combustion creates a failure spiral resulting in accelerated soot production and diminished airflow through the burner blast tube.

This downstream blockage will also lead to overheating of the oil burner blast tube and diffuser head (Figure 6-48), and in gas-fired appliances it will lead to overheating in the combustion chamber firebox area. Part of your preventative maintenance actions should include a visual examination of these components to make sure that they have not been damaged previously.

FIGURE 6-48 Overheated Burner Blast Tube and Diffuser Head

CHAPTER 7

THE GAS BURNER SYSTEM

Heating technicians who service appliances that burn natural gas or propane will encounter a wide array of equipment that will differ in age, design, and maintenance requirements.

Technicians should expect to confront:

- Boilers and furnaces that are brand new and older installations as well.
- Different appliances that are built with different burners that provide flame geometries appropriate for a particular appliance. Additionally, the design of the gas burner and ignition system is unique for that appliance.

The takeaway here is that technicians must be able to perform preventative and reactive maintenance (if possible) both on older equipment with scarcer or even nonexistent replacement parts and on newer, modern appliances of a more advanced design.

The Air-Gas Mix

Knowing the design of the appliance's construction is important in providing fast, efficient, and effective service repairs. Most residential heating gas burners are built using one of two basic designs, the *premix* and the *atmospheric tube* type burner.

The *premix burner* (Figure 7-1) is more common in modern higher-efficiency appliances. In this type of burner, the air and gas are mixed upstream of the appliance's combustion chamber prior to ignition. The mixture occurs outside the combustion chamber in a venturi mixer, and the burner has the ability to regulate the flame size and output.

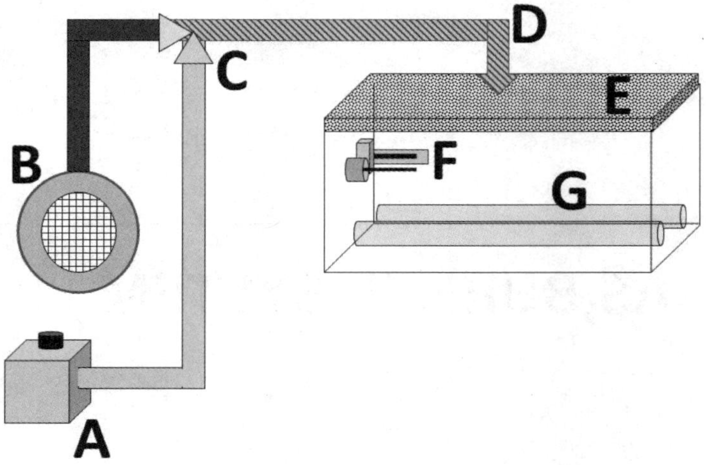

FIGURE 7-1 The Premix Combustion Process
(A) Gas Valve, (B) Variable Speed Fan, (C) Gas Admission Valve, (D) Air Fuel Mix, (E) Metal Fiber Diffuser, (F) Igniter and Flame Rod, (G) Heat Exchanger

In the more conventional *atmospheric or tube type of burner*, air and gas are supplied separately and mixed at the burner. A variety of atmospheric burners are available to provide a corresponding flame geometry appropriate for the appliance's combustion chamber. The example shown in Figure 7-2 is an atmospheric gas burner.

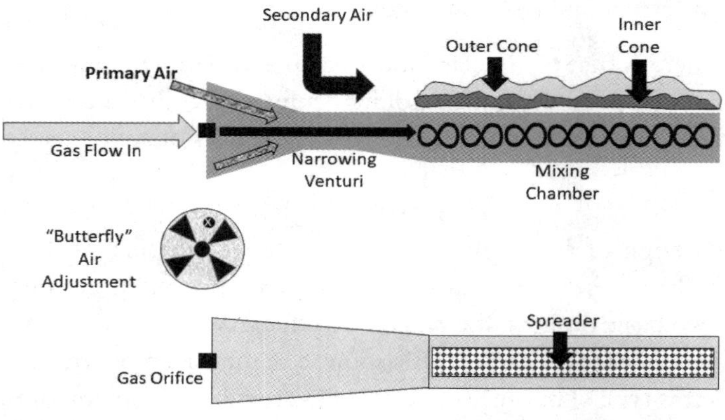

FIGURE 7-2 An Atmospheric Gas Burner

Ignition Systems

Knowing the design of the ignition system is just as important as knowing the design of the appliance's construction.

What all ignition systems must avoid at all costs is the flow and accumulation of unburnt gas, the consequences of which are obvious! Accumulated natural and other fuel gases are an asphyxiant and reduce normal oxygen concentration in the air that people breathe. This will lead to injury or death depending on the amount of fuel gas in the air.

Furthermore, an accumulation of fuel gases (whether it is natural gas or propane), given the right concentrations and an ignition source, will cause explosions and fires.

In short, the gas ignition system must function properly to prevent injury, death, and damage to the appliance or the property.

The Standing Pilot System

Appliances with the standing gas pilot ignition system are no longer being manufactured. But because this type of system was installed for many years, it is the most common gas ignition system that the heating technician will likely encounter.

The standing pilot system relies on a thermocouple (Figure 7-3) connected to the gas valve and positioned adjacent to a small "starter" (pilot) flame. When heated, the thermocouple generates millivolts that "allow" the main gas valve to open in response to a call for heat. Without the pilot flame present, the main gas valve will not open (Figure 7-4).

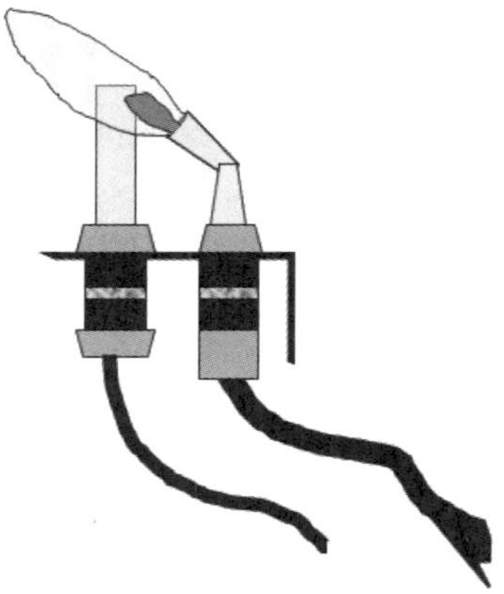

FIGURE 7-3 Standing Pilot Assembly

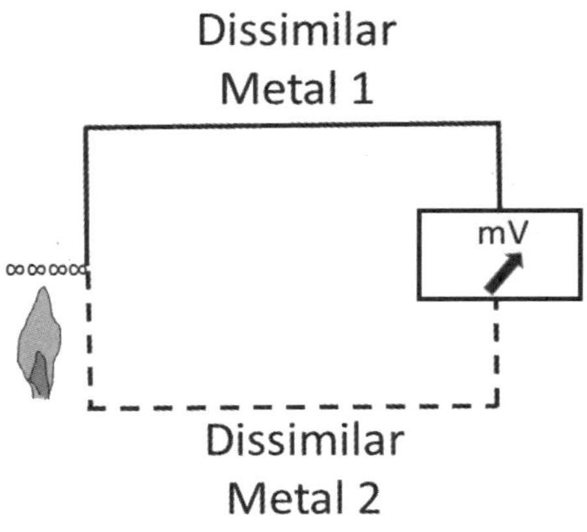

FIGURE 7-4 Thermocouple Design

The thermocouple and pilot assembly is positioned on the gas burner tube assembly (Figure 7-5) such that upon a call for heat, gas and air flowing into the burner tubes will be ignited quickly prior to a large accumulation of gas, preventing any delayed-ignition small "explosion."

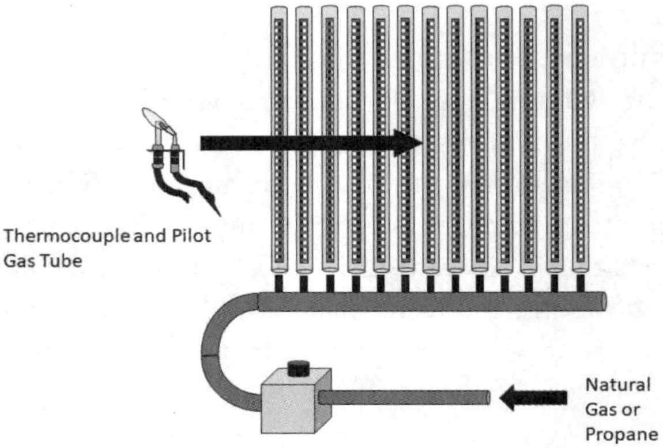

FIGURE 7-5 Thermocouple Gas Burner Tube Placement

As seen in Figure 7-6, the pilot flame assembly is positioned in the middle of the burner tube bundle to facilitate smooth gas combustion so as to centrally locate the ignition source in the middle of the combustion chamber.

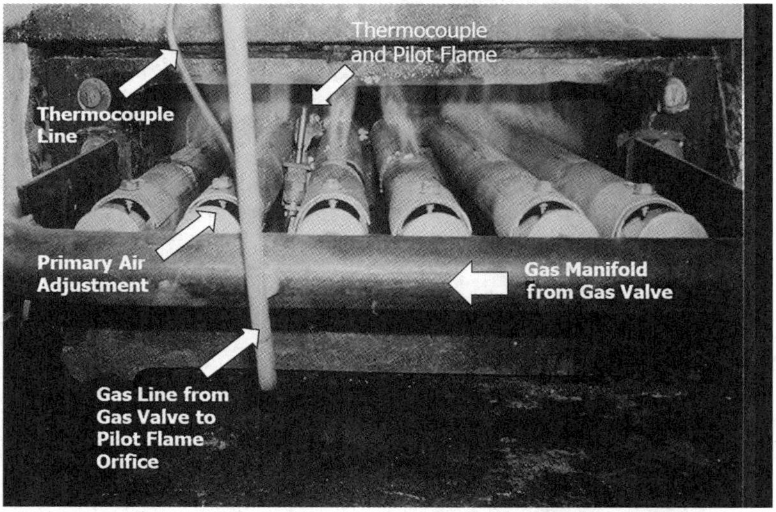

FIGURE 7-6 Pilot Flame Assembly in Operating Gas Burner

Energy Losses Due to Pilot Flame Usage

Energy losses from a pilot flame that uses gas on a continuous basis, coupled with the inconvenience and occasional malfunction of a pilot flame, have prompted manufacturers to design more reliable ignition systems for gas appliances.

Pilot flame sizes vary among different appliances. However, a typical usage rate can be identified by observing a gas meter in operation for a period of time. If, for example, it takes 93 minutes to consume 1,000 BTUs of gas, for 30 days this equates to 464,516.29 BTUs wasted. Converting BTUs to therms (where 1 therm is equivalent to 99,976 BTUs), the result is 4.64 therms used for the pilot flame per month. Multiply 4.64 therms by the actual cost of a therm of gas at $1.12, and you get a monthly cost of $5.19 for pilot flame losses for this appliance alone. For 12 months of pilot flame usage, this totals to $62.36.

And then there are other appliances in the household to consider, some with potentially larger pilot flames and hence greater usage of gas.

Using this model to estimate community-based usage for a city of 57,000 people living in 18,000 houses heated by gas, you get an estimated wasted gas cost between $1,122,480 and $1,683,720 just for pilot flame usage.

Electronic Ignition

Electronic ignition systems eliminate the inefficiencies of a continually lit, gas-consuming pilot flame by coordinating the flow of gas to the burner so that it coincides with gas flow to either the main burner in a *direct spark ignition system* or a pilot flame that only operates just prior to ignition in the *intermittent pilot system.*

An electronic control board or an ignition module is used to coordinate the initiation of gas flow and the igniter (Figure 7-7).

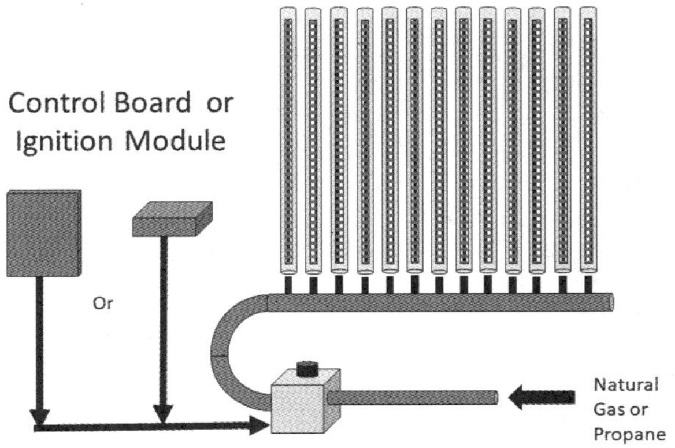

FIGURE 7-7 An Electronic Ignition Uses a Manufacturer's Control Board or an Aftermarket Ignition Module

Control Boards

Control boards (Figure 7-8) are supplied by the manufacturer as part of the furnace or boiler assembly. These are appliance-specific components that in the event of failure are not always easily replaced due to availability issues.

FIGURE 7-8 Furnace Control Board

Ignition Modules

Ignition modules (Figure 7-9) accomplish the same coordination of the gas flow and igniter operation as the manufacturer-installed control board. Some boilers and furnaces are outfitted directly from the manufacturer with ignition modules as opposed to control boards.

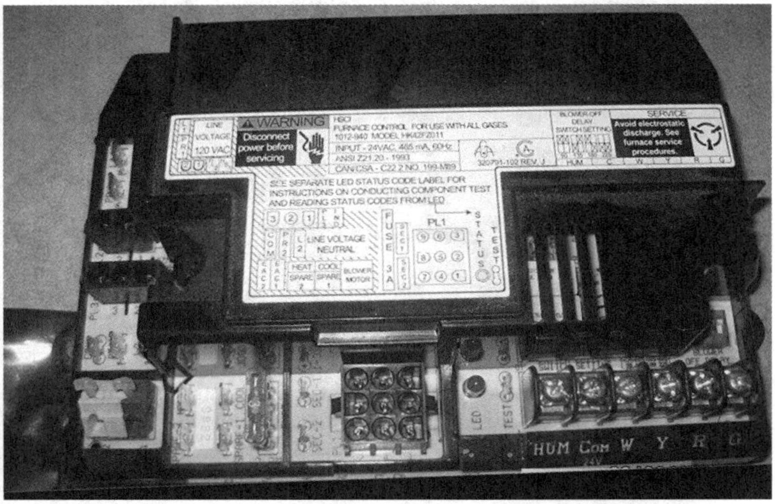

FIGURE 7-9 Ignition Module

An advantage of the aftermarket installation of the ignition module system is the ability to convert older standing pilot flame systems to intermittent pilot ignition systems, eliminating the need to relight extinguished pilot flames and reducing energy inefficiencies.

Another advantage of the ignition module is its ability to provide diagnostic codes indicating failures that may have caused the appliance to shut down. These can be especially helpful in intermittent failures where the shutdown may not be easily repeatable.

Flame Rectification

As is the case with all combustion controls, at all costs *no* fuel (gas) must accumulate prior to ignition. The verification of a flame being present at the igniter or burner, thereby eliminating the dangerous accumulation of unburnt fuel, is accomplished by the flame rod or flame rectifier.

The flame is rectified with the use of a flame sensor located in the pilot or in the burner flame for direct spark ignition systems (Figure 7-10). A current supplied to the sensor flows through the flame to the pilot assembly or burner head and then to ground.

FIGURE 7-10 Flame Sensor

Alternating current supplied to the flame sensor flows in one direction, which is rectified to a pulsating direct current. The direct current verifies to the control module that a flame is present, allowing the system to operate as long as the direct current at the flame rod exists.

If this direct current stops or falls to a low level, the control module will close the main gas valve, shutting down the system. To turn the system back into operation, the system may have to be recycled (shut off and on) by hand or may recycle itself automatically for a specific number of attempts to refire and return to operation.

Igniters

In lieu of a conventional standing pilot flame to ignite the gas supplied by the main gas valve, the gas being supplied to the pilot or main burner in an intermittent or direct spark ignition system is ignited with either a spark or hot surface igniter (HSI).

Hot Surface Ignition Hot surface igniters are silicon carbide or silicon nitride heating elements that glow red hot when electric power is applied to them. HSIs come in a variety of shapes and lengths appropriate for the operating electrical conditions, chamber configuration, and gas supply of the appliance (Figure 7-11).

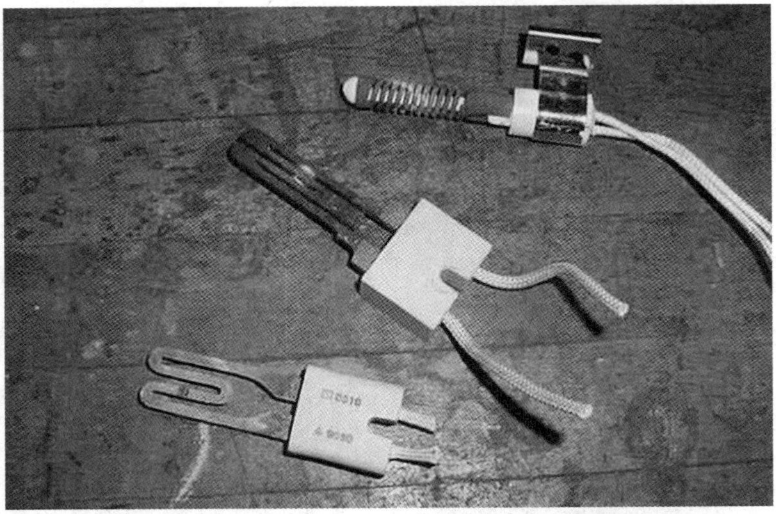

FIGURE 7-11 Hot Surface Igniters

The sequence of combustion may vary depending on the control board or ignition module being used. Typically, however, in response to a call for heat, the hot surface igniter will be energized, causing it to glow. This is followed by the main gas valve or gas to a pilot flame assembly opening allowing the flow of gas. This gas is then ignited when it comes into contact with the HSI-energized element.

A flame sensor adjacent to the igniter in the burner assembly (Figure 7-12) verifies that a flame is present, allowing the main gas valve to remain open and combustion to continue. At this point, voltage stops being supplied to the hot surface igniter. In the event that a flame is not sensed, another ignition initiation will be attempted. This cycle may repeat itself and, depending on the equipment being used, may do so up to four attempts.

The flame sensor–HSI system can be used to ignite a pure gas stream from the gas pilot valve or a gas-air mixture, as seen in the high-efficiency premix gas burner pictured in Figure 7-13.

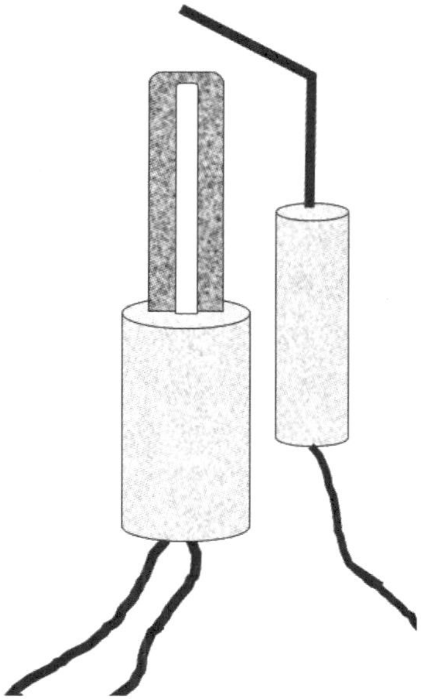

FIGURE 7-12 Flame Sensor and Igniter

FIGURE 7-13 Igniter and Flame Sensor in Premix Burner's Combustion Chamber

MAINTENANCE TIP!

Preventative Maintenance of Hot Surface Igniters

A hot surface igniter is a consumable item. Over time the effects of expansion and contraction due to continuous cycles of operation will cause it to fail (Figure 7-14).

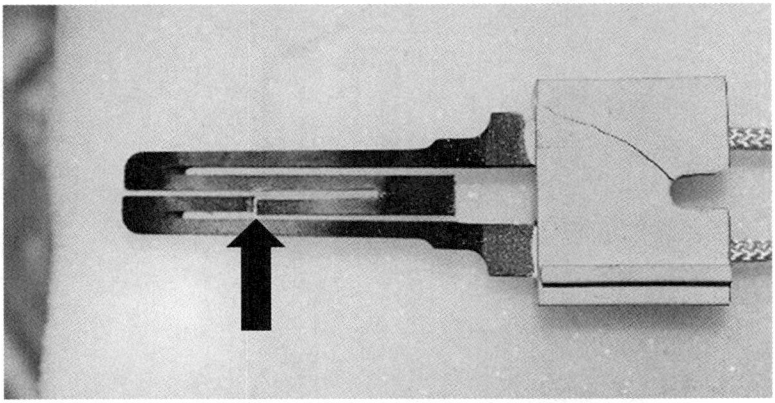

FIGURE 7-14 Failed Hot Surface Igniter

Igniters may last as little as 2 to 3 years or as long as 15. However, due to the relatively inexpensive cost of these components and especially in seasonal homes, proactive replacement of older igniters during preseason preventative maintenance should be considered.

The fragile nature of the HSI element may discourage the technician from visually examining it since the technician would have to remove the HSI before examining it and then put it back, and either action might cause damage.

The predictability of HSI failure is difficult since element failure may occur within the element itself or at opposite sides from the visible access points. That said, white residue on the element may be an indication of condensate contact or the vaporization of the silicon nitride or silicon carbide element.

Intermittent Ignition In the intermittent spark ignition system, when there is a call for heat from the room thermostat, the ignition control (control board or ignition module) is energized, lighting the pilot

This causes the pilot valve (which is inside the gas valve) to open and the spark igniter to ignite the pilot flame. Following this, the flame sensor (flame rod), as seen in Figure 7-15, verifies the presence of the pilot flame. At this point, the ignition control de-energizes the spark to the pilot, while at the same time the main gas valve opens.

On some models, if the pilot does not light, the system will shut down totally (lock out) until it is recycled by being turned off and on at the service switch. Older systems that do not have a lockout ignition system will continue to provide a spark until the pilot is lit. When the fuel supplied by the main gas valve ignites, the system will operate until the thermostat is satisfied, at which point the main gas valve shuts off and the cycle repeats itself.

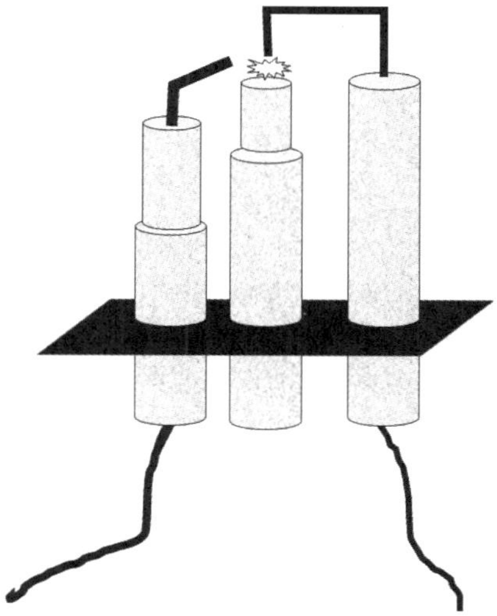

FIGURE 7-15 Spark Igniter and Flame Sensor

Direct Spark Ignition A rarer arrangement of an ignition system is the direct spark ignition. This type has no pilot light at all and lights the main burners on a call for heat. The system is outfitted with quick-acting flame sensors that shut down the main gas valve in a matter of seconds.

Mercury Switch Pilot Verification System The older, outdated mercury bulb–pilot switch verification system uses a mercury switch to verify the presence of a pilot flame. Mercury switch systems are present in older standing pilot flame appliances as well as in intermittent pilot flame ignition systems.

Once a pilot flame is established, the heat of the pilot flame is transferred to a mercury-filled bulb–capillary tube arrangement. The generated heat and pressure form the switch, which in turn allows the main gas valve to open. When the main gas valve opens, combustion is initiated in the main burners (Figure 7-16).

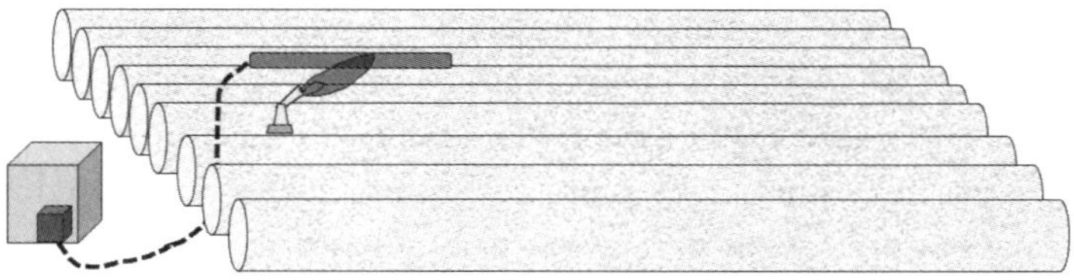

FIGURE 7-16 Mercury Switch Pilot System

The environmental hazards of mercury and the challenges of integrating these systems with modern appliances have resulted in a decrease in these systems' popularity.

Burner Design and Construction

As stated, gas burners are either *atmospheric-tube type* or *premix* designs.

Atmospheric or Tube Type

In the conventional, atmospheric, or tube type of burner design, combustion air and gas mix at the primary air interface where the gas nozzle meets the burner tube. An air adjustment (air shutter) may also be present here to make adjustments in the primary air addition to the combustion process. Additionally, at the outlet of the burner tube's flame spreaders, the burner flame receives secondary air. The gas burner delivers a flame of a consistent height, shape, and geometry to conform to the combustion chamber of the boiler or furnace (Figure 7-17).

FIGURE 7-17 Atmospheric Gas Burner Showing Air Shutters on each Burner Tube

Atmospheric burners are found in appliances that are 70, 80, and 90% efficient. With the exception of the in-shot burner type, these burners operate under air pressure and discharge their products of combustion into a chamber at atmospheric pressures. Note that the products of combustion are not mechanically drawn from the appliance by an induction fan, as they are in the inshot burner described below. Atmospheric burners are oriented horizontally, resulting in a vertical burner flame.

The burners themselves can be constructed of cast iron (Figure 7-18) or stamped steel (Figure 7-19) in a number of shapes to produce the desired flame shape and orientation. An example of the different burner shapes that can be produced is the upshot star-flame-pattern burner (Figure 7-20) common in older boiler gas conversion retrofits and in conventional gas hot water heaters.

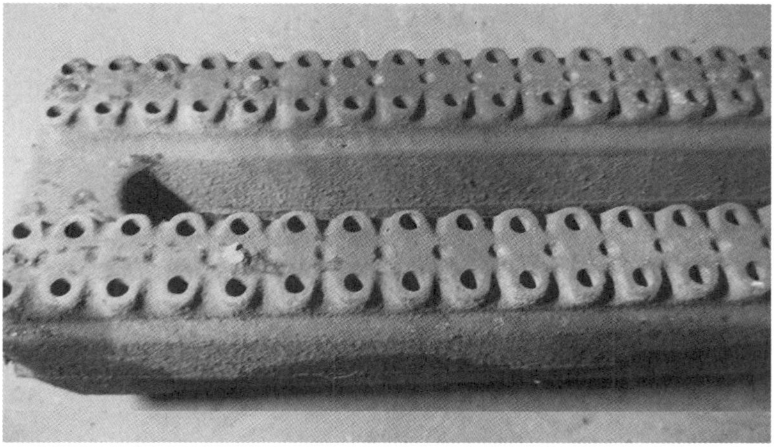

FIGURE 7-18 Cast Iron Burner

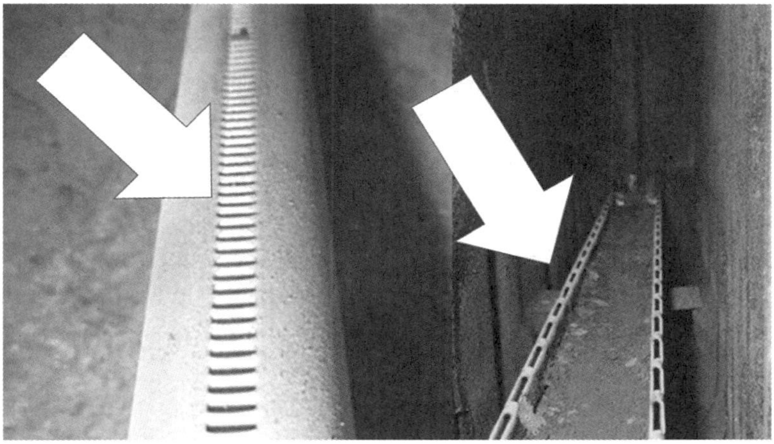

FIGURE 7-19 Stamped Steel Burner

FIGURE 7-20 Upshot Burner

Inshot Burner

Higher-efficiency appliances use inshot burners (Figure 7-21). Unlike atmospheric burners, inshot burners are subjected to the vacuum side of an induced draft fan that draws products of combustion from the appliance and in doing so helps to draw products of combustion from the burner.

FIGURE 7-21 Inshot Burner

Also unlike atmospheric burners where the flame is in a vertical orientation in the combustion chamber, the flame of the inshot burner is horizontally oriented, and there is no portion of the burner that is actually inside the combustion chamber of the appliance since the flame is literally "shot" into the appliance.

The Gas Conversion Burner

The gas conversion burner (Figure 7-22) is used on boilers or furnaces that have a design that allows, or that initially intended, the appliance to be outfitted with an oil burner. Since the oil burner produces a horizontally oriented flame, the conversion burner does the same, but it does so by burning gas or propane.

The gas conversion burner, much like the oil-fired burner, mixes air and fuel by utilizing a fan assembly, allowing the air-to-fuel mixture to flow down the burner's combustion tube or blast tube (Figure 7-23). This mixture is ignited by a hot surface or spark igniter just at the burner's exit and at the entrance into the boiler's or furnace's combustion chamber (Figure 7-24).

FIGURE 7-22 Gas Conversion Burner

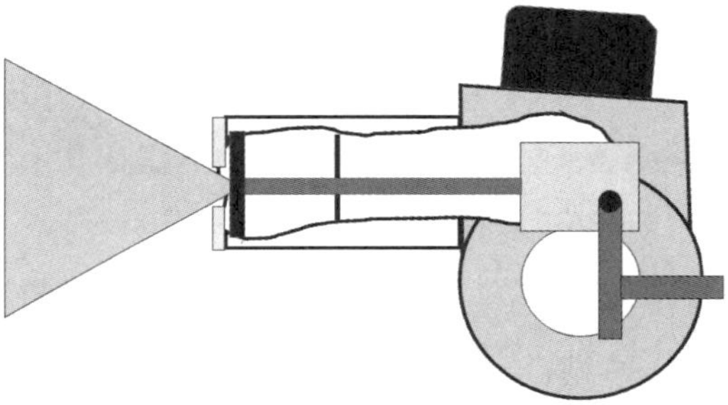

FIGURE 7-23 Cutaway of Gas Conversion Components

FIGURE 7-24 Conversion Burner's Igniter

Premix

As stated, premix burners mix gas and air in a venturi mixer and proportionately deliver this mixture to the appliance's combustion chamber prior to ignition.

The gas-air mixture is ignited in the chamber, usually through a metal fiber gas burner that equalizes this airflow to produce a uniform flow pattern and flame consistency. Depending on the appliance's combustion chamber geometry, the metal fiber burner may be either tubular shaped (Figure 7-25) for spiral-tubed circular heat exchanger boilers or rectangular for similarly shaped heat exchangers (Figure 7-26).

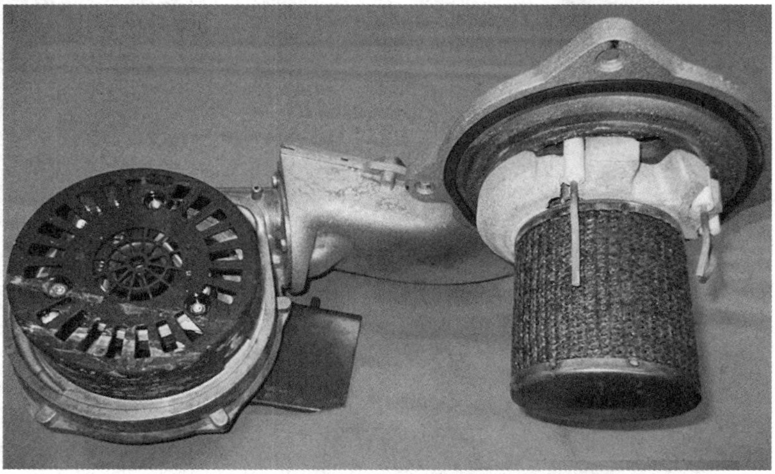

FIGURE 7-25 Tubular Premix Burner

FIGURE 7-26 Rectangular Premix Burner

The premix burner may utilize a high-temperature-resistant iron-nickel and aluminum alloy mesh that creates a uniform flame geometry appropriate for the combustion chamber of the appliance it is installed in.

Metal fiber gas burners (Figure 7-27) have the durability to function over long periods of time exposed to high temperatures without heat-induced oxidation causing them to fail and disturb the intended flame shape.

FIGURE 7-27 Metal Fiber Diffuser in Premix Burner

Gas Burner Maintenance

Many homeowners assume that gas-fired appliances do not require preventative maintenance although this assumption is incorrect. As is the case with any boiler or furnace, manufacturers' installation and operation manuals will list preventative maintenance actions and frequencies. Note that the typical time period of preventative maintenance periods is usually based on a yearly maintenance interval. That said, however, if the operating environment or frequent repairs dictate otherwise, maintenance intervals may need to be increased.

Appliance Maintenance

In addition to maintenance of the burner, the overall condition of the appliance that the burner is connected to must be evaluated to look for and proactively correct any observed deficiencies or signs of impending failures. Items to search for include dried and cracking or leaking gaskets on circulator pump flanges and should be corrected if found. Signs of water leakage due to either backing up or leaking condensate, to leaking boiler pipe fittings, or to leaking section connections should also be an area of focus.

Where possible, limit switches and safety controls should be tested to determine if they are in good working order and capable of stopping the operation of the appliance in the event of an unsafe condition such as a flameout or overheating.

Gas Pressure Check

The first step in maintaining the gas burner, though, is to measure the gas pressure being supplied to the appliance to verify that the pressure is in compliance with the manufacturer's instructions and that it has not changed due to any regulator malfunctions. Depending on the system and equipment, gas pressure is measured at the gas valve (see the arrow in Figure 7-28) and the manometer connection to the gas valve port (see the arrow in Figure 7-29) using a pressure gauge or manometer (electronic or manual). Gas pressure is usually measured in inches of water column (Figure 7-30),

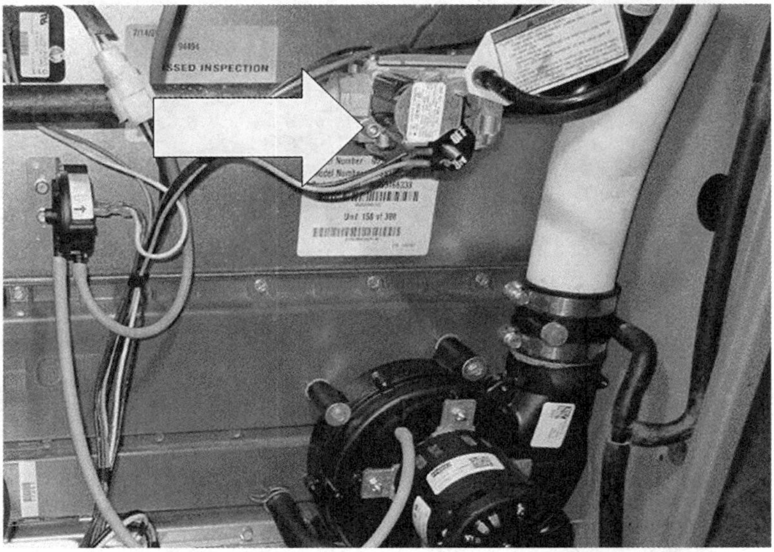

FIGURE 7-28 Measuring Gas Pressure at Gas Valve

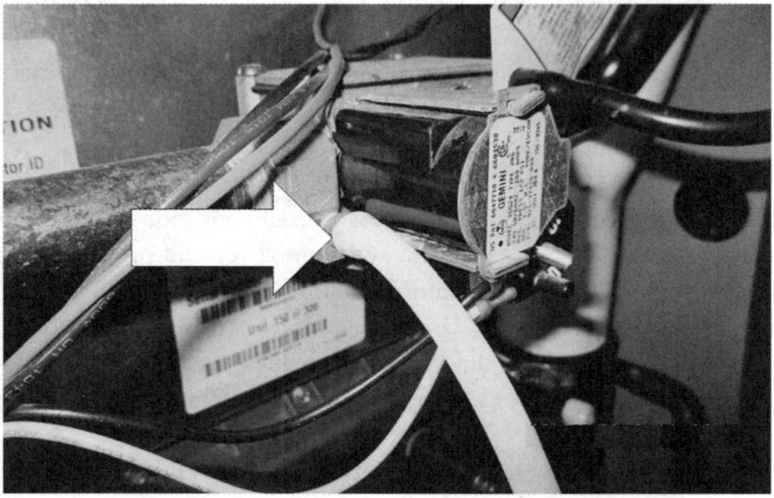

FIGURE 7-29 Manometer Connection to Gas Port

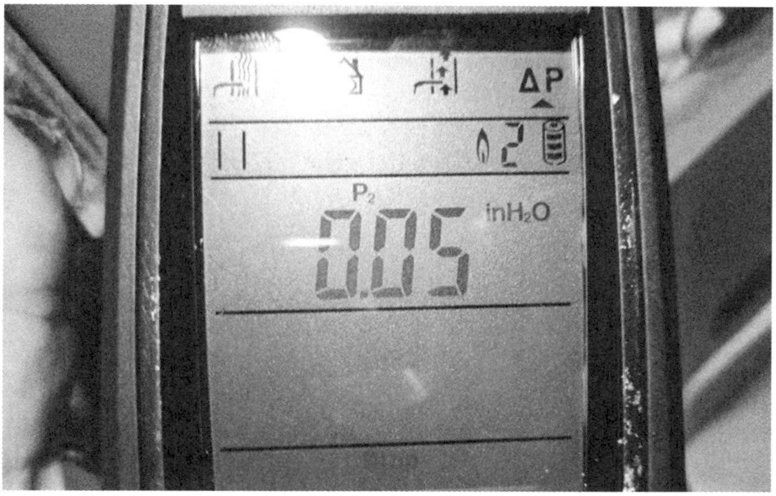

FIGURE 7-30 Analyzer (Courtesy of Testo)

Burner Tube Inspection

An often overlooked and easily remedied problem with gas burners (Figure 7-31) is their failure to operate properly. These components of the burner system are often the cause of problems such as soot production and carbon monoxide spillage as well as flame reversal (rollout) and delayed-ignition explosions.

FIGURE 7-31 Products of Combustion Backing Out of the Appliance Also Known as Rollout

During annual service events the burners should be removed from the appliance and thoroughly inspected to make sure that they are clear of the dust and debris that the burners draw in as they ingest air. Dust, pet hair, lint, and other airborne contaminants often accumulate at the primary air intake (Figure 7-32) as well as inside the burner tube. These substances will reduce the quantity of air that is available for combustion, causing the burner to produce soot.

FIGURE 7-32 Dust-Fouled Gas Burner Tubes

As well, debris on top of the burner tube, such as failed refractory pieces, will fall on the burner tube "flame spreader" openings, causing uneven gas distribution, flame impingement, and flame height distortion, causing soot production.

Also, corrosion (Figure 7-33) or heat damage to burner tubes will cause uneven flow of gas and possibly even obstruct the flow through the flame spreader enough so that no or partial combustion of fuel gas occurs. This in turn can cause the reversal of burner flame by-products as well as carbon monoxide spillage into the interior of the home.

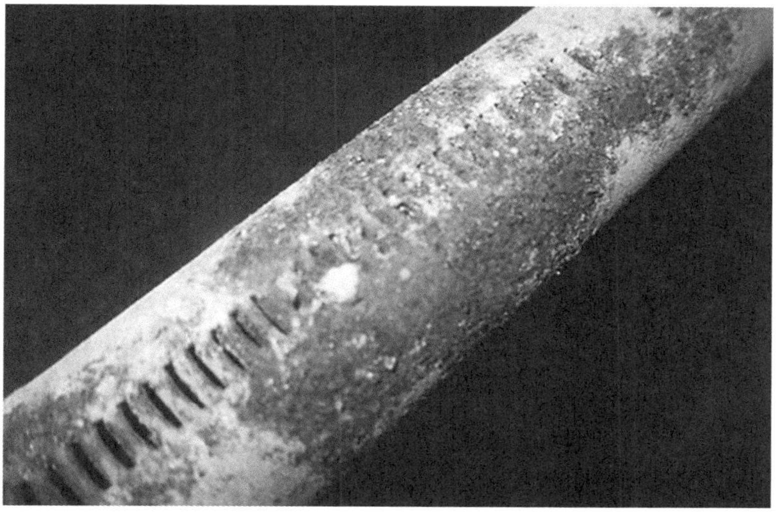

FIGURE 7-33 Burner Tube Flame Spreaders Obstructed by Corrosion

Combustion Chamber and Refractories

Removal of the gas burners also facilitates a thorough inspection of the furnace's heat exchanger and boiler combustion chamber area. Without removing the burners for cleaning, these areas of the appliance would be difficult to access for preventative maintenance inspection and proactive repair or appliance replacement, necessary steps to avoid a more serious failure occurring.

Since the heat exchanger and boiler combustion chamber are the areas where combustion occurs, they are most vulnerable to heat damage and as such are a logical area to check. In addition, combustion chamber refractory panel damage and soot production due to burner malfunction may be easily overlooked unless the gas burners are removed, allowing a clear line of sight to these areas. Damaged combustion chamber refractory panels and resulting soot production (Figure 7-34) can cause carbon monoxide spillage and appliance damage that can result in fires and injury to the occupants of the home. Likewise, failed or fallen refractory panels (Figure 7-35) can result in hot combustion gas leakage and damage to the firebox.

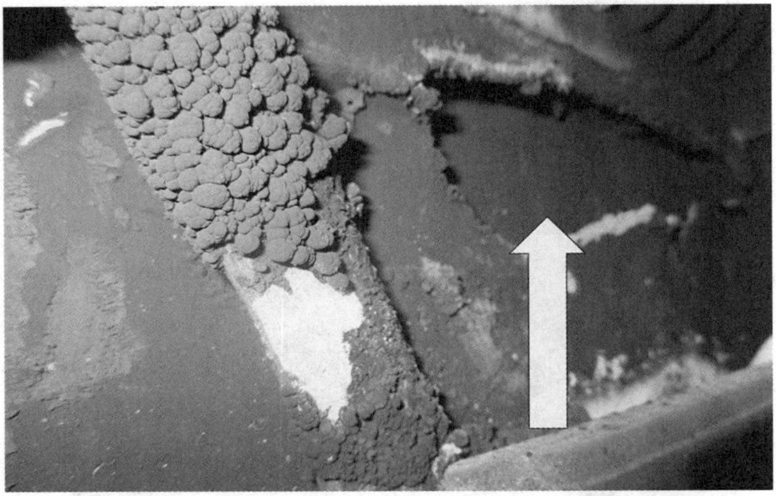

FIGURE 7-34 Damage to Combustion Chamber Refractory Panels Have Interfered With Burner Operation Causing Soot Production

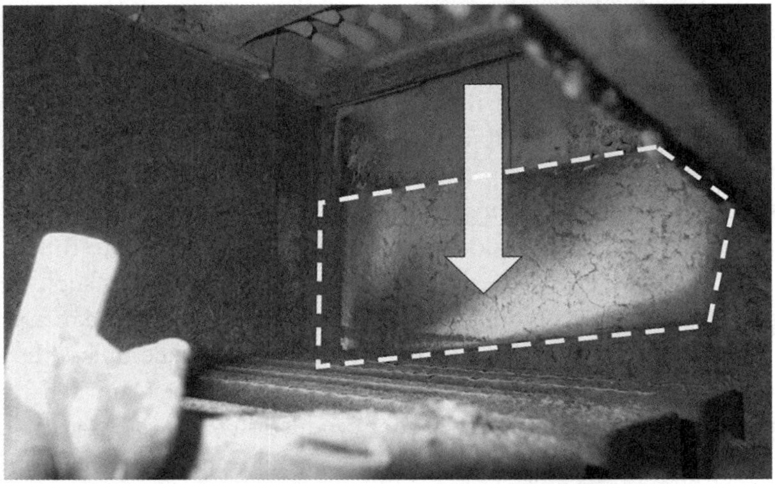

FIGURE 7-35 Fallen Combustion Chamber Refractory Panel in Gas-Fired Boiler

Combustion Performance Testing—Gas-Fired Appliances

To reiterate, you cannot conclusively tell how safely, efficiently, or correctly a burner (oil, natural gas, or LP gas) is operating by adjusting the flame by visual observations alone. Just as with oil-fired appliances, a combustion performance test (which is essentially a series of tests that assess the overall efficiency of the appliance) must be performed. For gas-fired appliances that appear to be operating correctly, at the beginning of a servicing event, it is a good time to perform a combustion performance test to establish a baseline reading of how it's performing prior to any maintenance being done. This information can be compared to combustion performance data at the end of the service event for comparison.

The first step is to make sure that the boiler or furnace has adequate draft, airflow, or water flow across the heat exchanger or boiler block and that fuel pressures are within manufacturers guidelines. This requires that fuel pressure be measured at the burner supply as well as draft through the heat exchanger be established initially and set to the manufacturer's recommendations.

Since firebox leakage can be affected by airflow, sheet metal jacket covers, rollout shields, and any other appliance component should be installed so that you are testing the appliance in the most representative "run configuration," just as it will be when in operation.

These gross basic overall operational parameters must be verified first before combustion testing is begun and the resulting burner adjustments are made.

Carbon Monoxide in Combustion Gases

Inadequate oxygen in a combustion process prevents the appliance from converting the hydrocarbon molecule of the fuel into the desired carbon dioxide gas. This is usually the result of inadequate combustion air or any of the numerous appliance malfunctions that can occur that affect the air-to-fuel ratio operational parameters of the burner. With inadequate combustion, the imbalance of the air-to-fuel ratio allows the formation of carbon monoxide, a poisonous and odorless gas that interferes with the ability of oxygen to be carried in the bloodstream. This can result in dizziness, headaches, nausea, and death to the building's occupants.

Although complete conversion of the fuel to 100% carbon dioxide is desirable, even for a properly running appliance some levels of carbon monoxide result. This is acceptable if (and only if) these levels are low and the gas is inside the appliance firesides and chimney or exhaust venting piping.

Ambient Carbon Monoxide Measurement and Levels

An ambient carbon monoxide measurement (Figure 7-36) identifies the carbon monoxide outside the appliance. Conditions such as product-of-combustion spillage due to a damaged or malfunctioning gas burner, an obstructed appliance flue passageway, or chimney blockage will cause products of combustion to back up and leak from the appliance.

When a CO analyzer is used in the airstream of a gas-fired furnace (preferably as close to the downstream airflow from the heat exchanger as possible), it can identify breaks or cracks in the heat exchanger as well. When increasing levels of carbon monoxide or sustained levels that are contrary to the manufacturer's recommendations are detected, the technician will need to implement a more aggressive investigation into the cause of the appliance's conditions.

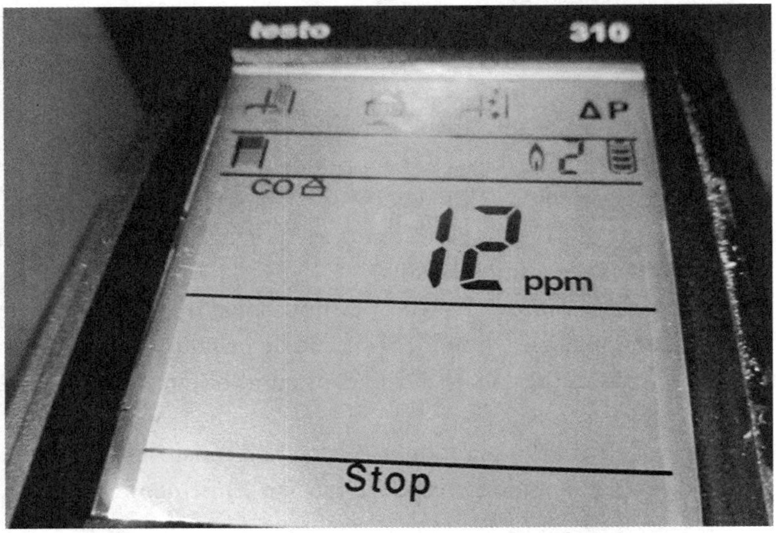

FIGURE 7-36 Ambient Air Testing Combustion Analyzer (Courtesy of Testo)

Although no level of carbon monoxide is desirable, some carbon monoxide may be present in a home due to cigarette smoke or emissions from appliances such as gas-fired cooking ranges or fireplaces. Maximum safe carbon monoxide levels are set by local jurisdictions that have code enforcement authority.

In regard to symptoms of CO poisoning, according to the United States Consumer Products Safety Commission, people exposed to carbon monoxide levels between 1 and 70 ppm may not exhibit any symptoms. For levels above 70 ppm, people can experience headaches, fatigue, and nausea; and between 150 to 200 ppm, disorientation, unconsciousness, and death are possible.

Appliance Combustion Parameters

As part of combustion performance testing of gas-fired appliances, technicians need to measure:

- Draft
- Oxygen
- Carbon dioxide
- Carbon monoxide
- Flue gas temperature
- Ambient temperature
- Stack temperature

(Smoke measurement testing is not done on gas-fired appliances.)

To measure these combustion parameters, technicians can use either individual electronic combustion performance testing devices or a combination of individual instruments (such as a manometer and thermometer). The final combustion efficiency is calculated from the values of the measured readings either automatically as is the case in electronic analyzers or from tables provided with conventional Fyrite-based equipment.

Modern combustion analyzers integrate many needed functions of combustion performance testing. These instruments simultaneously measure combustion parameters and print the data (Figure 7-37) to show the homeowner the operating conditions of the appliance as well as provide a record of its performance on that date to future service technicians.

It is important for technicians to document on paper that they carried out a combustion performance test initially as well as during any subsequent service call on the appliance. By providing a copy to the homeowner and retaining a copy of the testing for themselves, they can prove that they did everything in their power to make sure the appliance was operating in an acceptable manner at the time of the service call.

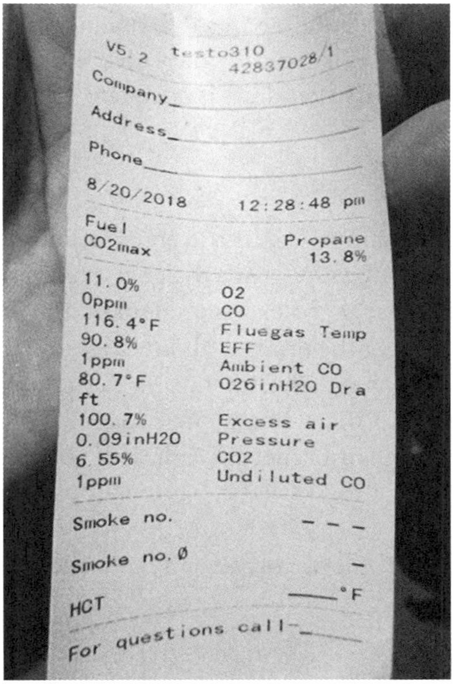

FIGURE 7-37 Combustion Performance Result Slip Left on the Appliance

The following discusses the individual combustion performance test measurement parameters, and Table 7-1 presents typical gas-fired appliance setup readings for the parameters.

TABLE 7-1. Typical Gas-Fired Appliance Setup Readings

Parameter	Atmospheric Draft, 75–80% Efficient	Gas-Fired Power Burners	Draft Sealed Combustion Induced Draft, 88–92% Efficient	Draft-Induced Gas Burners, 80–82% Efficient
O_2	7–9%	3–6 %	5–7%	7–9%
CO_2	6.5–8%	8.5–11%	7–8.5%	6.5–8%
Stack temp	325°–500°F	320°–570°F	<125°F	325°–400°F
Draft	0.02–0.04-inch water column	0.02–0.04-inch water column (over the fire)	0.02–0.008-inch water column	0.02–0.04-inch water column
CO (undiluted)	<100 ppm	<100 ppm	<100 ppm	<100 ppm

Draft

The flow of combustion gas through the firesides of the boiler or furnace is the first needed operating parameter of the appliance's operation. This establishes the appliance's ability to receive combustion gases from the burner and remove them from the appliance to the exterior of the building in a safe manner.

In the United States, draft is typically measured in inches of water column, and appliances are designed to verify that draft is adequate and in a direction as intended by the manufacturer.

O_2

By percentage, air contains 20.9% O_2. In the combustion process, air is mixed with fuel to produce complete combustion, with the goal of eliminating the production of carbon monoxide. To achieve this, an excess amount of air is supplied to the combustion process either by adjusting the primary air intake shutters for (older) atmospheric gas-fired burners or by making air adjustments. Modern furnaces may not have an air adjustment incorporated into their design since manufacturers designed these with excess air incorporated into their burners' design with no need of adjustment.

CO_2

Each fuel has a maximum CO_2 level corresponding to the fuel that the appliance uses by volume:

- No. 2 fuel oil: 15.4%
- Natural gas: 11.8%
- Propane: 13.8%

Carbon dioxide is the main indicator of the efficiency of the appliance's combustion process. A high CO_2 reading with a low excess air level results in more of the heat from the combustion process being used for useful (heating) purposes and less heat being lost.

CO

When measuring carbon monoxide levels in the fireside–internal combustion gases within the appliance, remember that these will not result in steady-state readings from the initiation of combustion to the end of the heating cycle. High CO levels at the initiation of combustion light-off are to be expected. However, abnormally high levels that do not drop to the manufacturer's acceptable levels after 5–7 minutes of operation require further examination to determine the cause.

To identify the operational steady-state CO levels, monitor the CO content of the combustion gases throughout an operational cycle, making note of the decrease in the CO levels and the point to which they fall and at which they remain.

Ambient Air Temperature

The ambient air temperature is the temperature of the air surrounding the appliance. This temperature is used to adjust the net temperature increase across the appliance, which is used to determine the appliance's overall efficiency. The ideal location to measure the ambient temperature is at the combustion air inlet area since other areas directly adjacent to the appliance may be warmer than this, affecting the overall efficiency calculation.

Stack Temperature

The stack temperature is measured at the point in the exhaust connector pipe upstream of any barometric or other air damper in the connector pipe of a boiler or furnace. In metal breaching connector pipes, a probe is inserted through a hole (usually about ¼ inch in diameter) in the metal stovepipe or in the PVC discharge piping in condensing high-efficiency appliances in either a provided sampling port or a drilled hole approximately 18 inches from the appliance exit. The sampling hole can be sealed with a tapered plug or screw.

The stack temperature of the combustion gases is the hottest temperature that the appliance will produce and is the point on the combustion stream where the CO_2 level is at its highest and the O_2 level is at its lowest. The stack temperature must be high enough to prevent condensation in standard 80% efficient noncondensing appliances and low enough in high 90%-plus efficient appliances to make sure enough heat is being removed from the furnace's or boiler's heat exchanger as intended.

In closing, it is true that gas-fired appliances may operate with less soot production than oil-fired appliances, and they require less stringent preventative maintenance actions because there is no need to replace filters and nozzles annually. But a scheduled maintenance program is still needed. By implementing a maintenance program on a regular basis, potentially hazardous conditions such as carbon monoxide spillage and progressing equipment damage can be identified and corrected before the appliance fails to a more damaging and catastrophic degree.

Remember, the manufacturer's manual and your experience with similar appliances are your guide for preventative maintenance scheduling. It is your job to know where to find the information needed to not only install but also maintain the appliance throughout its lifetime. In the event that you cannot ensure that the appliance is running in a safe manner or that you cannot be sure the appliance will run safely until the next scheduled maintenance appointment, it is your duty to do something about it. In the end, the homeowner is relying on your judgment and experience to make sure that the heating system will operate as intended.

References

Bergmann, James L., *Combustion Analysis: An Essential Reference for the Advanced Technician*, Testo, Inc., Flanders, NJ, 2005.

TSI Incorporated, *Combustion Analysis Basics: An Overview of Measurements, Methods and Calculations Used in Combustion Analysis*, 2004.

CHAPTER 8

SOOT DAMAGE AND POOR INDOOR AIR QUALITY IN HOMES

Indoor Air Quality

Negatively affecting the quality of indoor air, pollutants ranging from mold and dust to soot and other dirt have become a nuisance and, in some instances, a health issue that too many homeowners have to suffer (Figure 8-1).

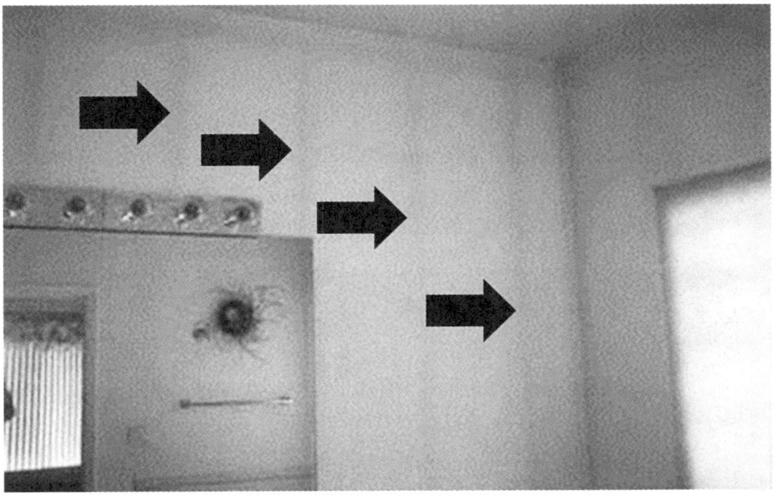

FIGURE 8-1 Deposits on Interior Wall Surfaces Outlining Wall Framing Studs

Sometimes it is easy to trace the cause of the pollutants. For example, most of the time the cause of a "puffback," or soot discharge into a home, is known. However, many times homeowners—and technicians as well—do not know the origin of these pollutants. Likewise, what may appear to be the source may, in fact, not be. This results in expended time and resources in an effort to find the source as well as the cause of the damage.

Mold forms in homes where temperature and moisture conditions support the production of mold spores. Conditions such as roof leaks, improper ventilation, excessive moist air disposal, and leaking plumbing and heating pipes can introduce moisture into the building envelope, which leads to the growth of biological contaminants.

Likewise, the dust and dirt that form in heating system baseboard elements (Figure 8-2) and ductwork (Figure 8-3) can become airborne and eventually accumulate in concentrations that are then noticed visually or even by people's allergic reaction to them.

FIGURE 8-2 Dust Accumulation in Baseboard Element and on Cover of Forced Hot Water System

FIGURE 8-3 Forced Hot Air Ductwork Fouled with Accumulated Debris

FORENSIC INSIGHT

Identifying the Origin of the Deposit. In Doubt? Check It Out!

The naked eye cannot accurately identify the source of a "soot-like" or other poor indoor air quality deposit. Differentiating between wood, cigarette, or other combustion by-products, and hence being able to determine the activity or appliance that is the cause, cannot be done conclusively without laboratory assistance. Given a sample of a particulate, a laboratory can identify the source of that material scientifically by using the physical characteristics of the particulate and its chemical composition

One method is to identify the physical characteristics of the deposit using a stereomicroscope to examine the samples. This method focuses on the geometry and physical characteristics of the deposit to identify its origin. This is usually done by comparing the magnified appearance of the unknown substance with images of other samples of a known origin (Figure 8-4).

FIGURE 8-4 Magnified Soot Particles

Another method is to identify the types of elements in samples of the soot using a scanning electron microscope in conjunction with an energy-dispersive X-ray analysis (EDX) (Figure 8-5). This method analyzes the chemical composition of the samples. Once the elements are identified, they can be compared with the known chemical content of possible fuel sources to determine the exact source.

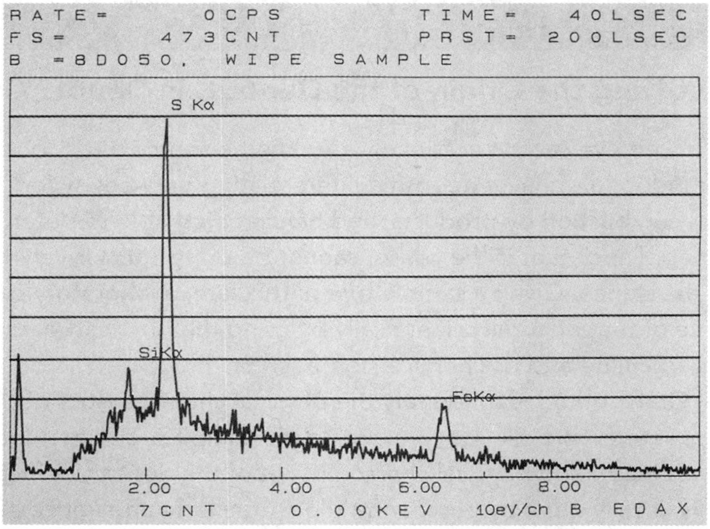

FIGURE 8-5 Energy-Dispersive X-ray Analysis of Soot Sample

It's Soot! What Is the Cause?

Say that it has been concluded that the contaminant in the home is, in fact, soot, a by-product of combustion that likely originated from the heating system. The next step is to investigate where and how the contaminants are leaving the confines of the appliance and exhaust system and finding their way into the building interior.

Conventional Failure Diagnosis

Often the cause of the leak is the burner or furnace, so the conventional place to start the investigation is with that equipment.

Malfunctions That Cause Rollout and Backpressure Product-of-Combustion Spillage

Malfunctions in boilers and furnaces can result from typical issues such as a burner malfunction that causes the appliance (Figure 8-6) to create soot. As this soot accumulates, the flue passageways of the appliance or even the chimney becomes obstructed, which causes the products of combustion that contain the soot to back out of the appliance and into the home (Figure 8-7).

FIGURE 8-6 Forced Hot Air Furnace's Heat Exchanger Obstructed with Soot

FIGURE 8-7 Soot Discharge from Furnace with Obstructed Flue Passageways

Another malfunction is a delayed ignition explosion (Figure 8-8). This will result in soot accumulations outside the appliance, and if the malfunction is left undetected, these particulates can accumulate over time and damage interior surfaces.

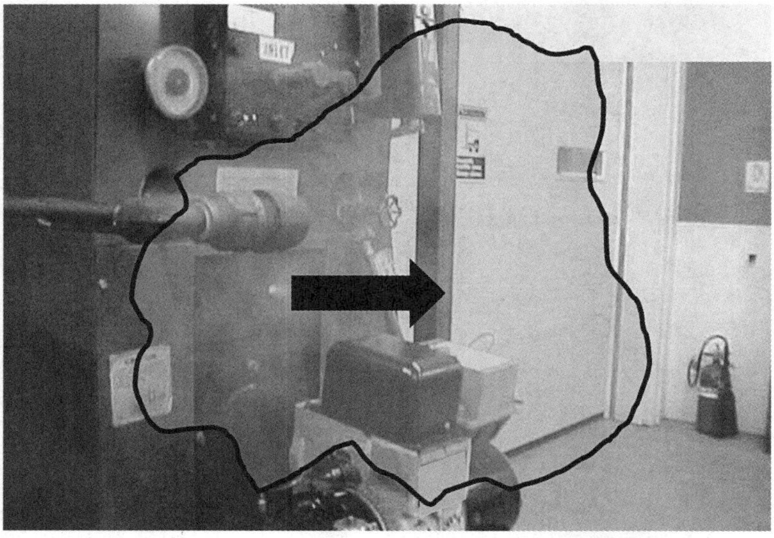

FIGURE 8-8 Delayed Ignition Explosion Soot Cloud

Using Gas Detectors and Other Electronic Equipment

The identification of products of combustion leaking directly into the interior of a home may be obvious (Figure 8-9). However, in smaller quantities or in intermittent circumstances, these products of combustion may be more difficult to identify. In these cases especially, combustion gas spillage conditions can be detected, oftentimes without disassembling the appliance, by the use of a product-of-combustion gas detector or an infrared imaging camera, as seen in Figure 8-10.

FIGURE 8-9 A backup of products of combustion in a gas- or oil-fired appliance will result in rollout or positive combustion chamber backpressure that will spill products of combustion into the interior environment when the appliance fires.

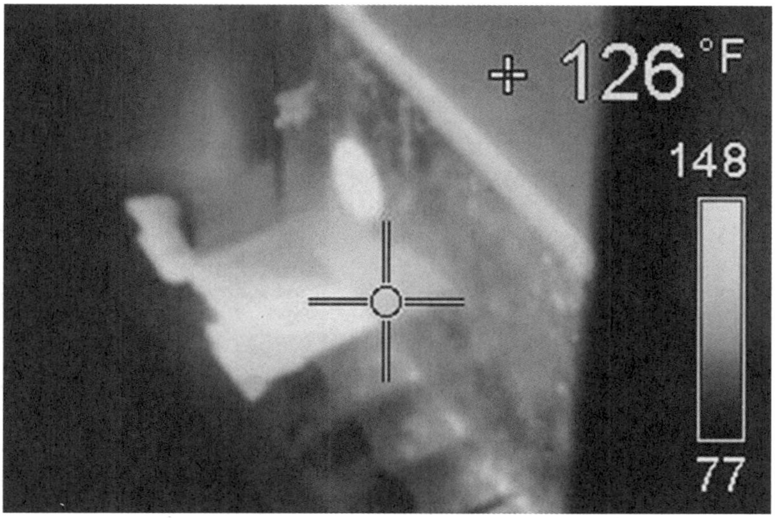

FIGURE 8-10 Infrared Heat Detection Cameras Are Helpful in Identifying Hidden Combustion Gas Leaks

Gas detectors and other electronic detection devices can be used in a number of investigations:

- Product-of-combustion gas detectors can be used to identify combustion gas spillage due to an obstructed combustion gas pathway.
- Product-of-combustion gas detectors can also be used on operating appliances to identify internal leakages under the boiler or furnace jacket. Conditions such as a leaking furnace or boiler cleanout openings (Figure 8-11) can be identified in this manner.

FIGURE 8-11 Product of Combustion Gas Detector Testing for Leakage at Furnace Cleanout Opening (Courtesy of BOSCH)

- A gas analyzer can be used to monitor the space between the heat exchanger and jacket to identify combustion gas leaking between boiler sections or from a combustion chamber (Figure 8-12) while the appliance is in operation.

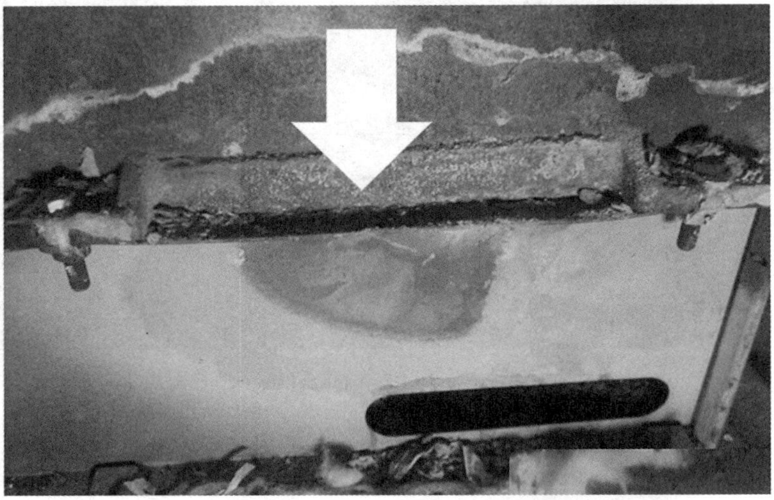

FIGURE 8-12 Combustion Gas Leak in Gas-Fired Boiler Originates at Interface Between Combustion Chamber and Boiler Block

- If access to a product-of-combustion gas detector is not possible, removal of the furnace or boiler jacket may also reveal soot staining (Figure 8-13) on the insulation between the heat exchanger and jacket, providing an indication of combustion gas leakage.

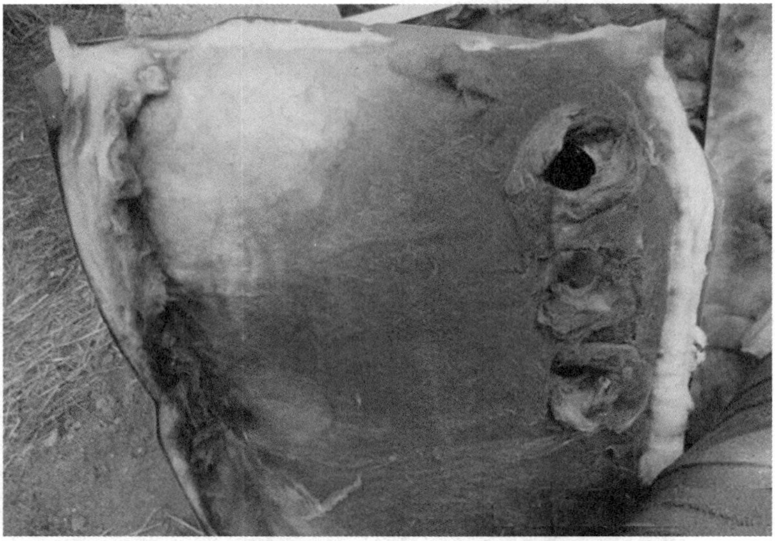

FIGURE 8-13 Removal of Furnace Jacket Identified Overheated and Soot Stained Insulation Indicative of a Combustion Gas Leak from the Appliance

For forced hot air furnaces, a gas analyzer monitoring the quality of the system airstream from a forced hot air furnace (Figure 8-14) can oftentimes provide information indicating that cracks, flame impingement holes, or other combustion gas leaks are present.

FIGURE 8-14 Testing the Supply Air from a Furnaces Hot Air System (Courtesy of BOSCH)

Access to the origins of product-of-combustion leakage is sometimes difficult without disassembling the appliance. However, the use of a product-of-combustion gas detector can provide an indication that there is combustion gas leak in an area within or emitting from the appliance (Figure 8-15).

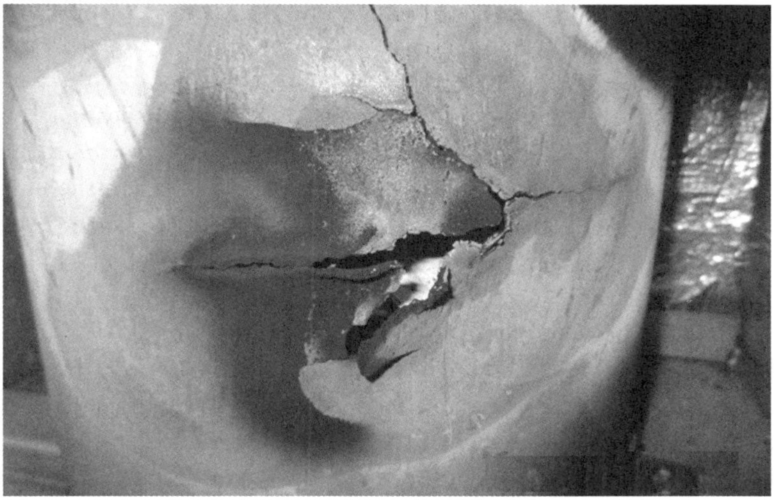

FIGURE 8-15 Investigation Identified the Presence of a Flame Impingement Hole in the Furnaces Heat Exchanger

Common Locations of Accumulated Soot

Regardless of where the contaminant originates, contaminants introduced into the interior of a structure will accumulate in common locations that in themselves are an indication of a poor indoor air quality issue.

Ghosting Soot Accumulations

Once inside the interior conditioned space, carbon particles from the combustion appliance source will accumulate in common locations. The streaks and spots they form are known as "ghosting soot." Due to interior air current flow, electrostatic forces, and temperature differences, ghosting soot is often found on surfaces such as plastic toys or cookware. Other common locations where ghosting soot accumulates are on the walls above forced hot water baseboards. See Figures 8-16 and 8-17.

FIGURE 8-16 Soot Particles Adhering to Wall Above Baseboard

As seen in Figure 8-17, convective air current flows in a room being heated by steam or hot water, the flow of air will follow a circular path with cold air entering the bottom of the baseboard element where it is heated. The warm air then rises to begin another turn through this heating cycle. Any particulates in the indoor air are entrained in the circulating air. As these particulates exit the baseboard, due to static electric charge, they will adhere to the wall just above the baseboard, and over time they will become noticeable.

These emission particle accumulations and deposits are also found on the parts of walls that cover the studs behind sheetrock, as verified through the insert infrared image photograph in Figure 8-18, showing the location of the framing members.

Framing members are colder than the insulated wall cavity. When moist indoor air comes in contact with the parts of the wall that cover the cooler wall studs, it condenses, allowing the air particulates to adhere to these locations.

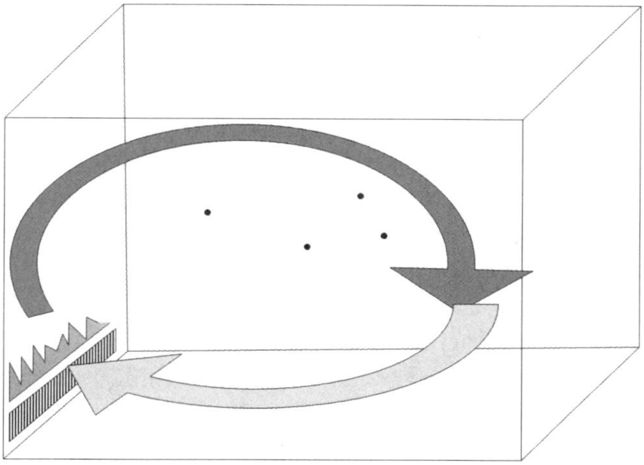

FIGURE 8-17 Particle Transport Due to Convective Air Current Flow

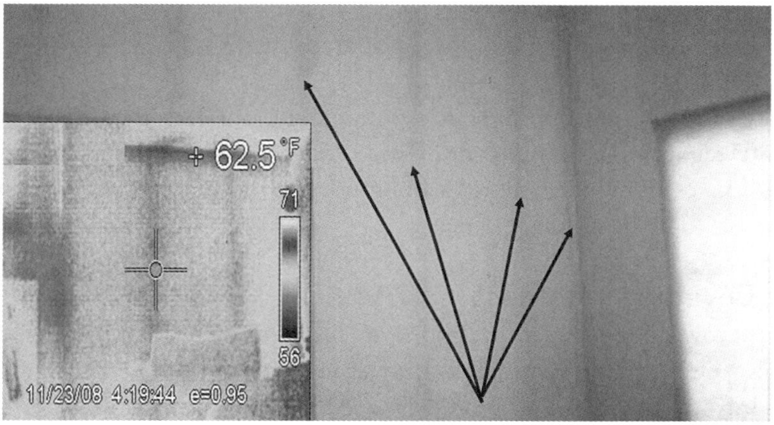

FIGURE 8-18 Soot Stains in Sheetrock Coincide with Locations of Framing Members

Depressurization-Induced Combustion Reentry

OK, so the boiler or furnace is *not* leaking products of combustion, is not backed up, does not have an obstructed chimney flue, and is operating as intended without malfunction or failure. That said, there *is* soot in the house.

Why is this happening?

With the advances in building materials and practices through the building expansions of the 1980s and 1990s, along with the inherent benefits of these improvements, some unwanted side effects have resulted as well.

One such chronic building issue is the result of the failure to recognize the house-heating (or combustion) appliance as a "system." This problem is known as *depressurization-induced combustion reentry.* Depressurization-induced combustion reentry is the reintroduction of combustion emissions into a structure as a result of negative air pressure developed by the structure as well as the location of the point of discharge of the emissions from the combustion appliance. The combustion appliance might be a boiler, furnace, or fireplace, and this problem occurs regardless of the type of combustion fuel being consumed in the structure.

Advancements in Housing Material

House construction in the United States increased significantly in the post–World War II building boom through the 1950s and early 1960s. Then, in the building boom of the 1980s and 1990s that accommodated the growing families of the baby-boom generation, advancements in the ability to increase the integrity of the standard housing building envelope allowed buildings to achieve thermal efficiencies far exceeding what was capable in the past. The impetus for thermal efficiency and a tighter building envelope was born out of the energy crisis in the mid-1970s, which also saw an increase in the use of alternative energy sources, such as wood-burning and solar heating systems.

At the same time that the advances in building materials developed, there was a "lag-behind" regarding building codes. These codes were unable to keep up with not only the advances in building materials but also the modifications, construction, and installation of home heating appliances.

Building Envelope Tightness

With an eye toward less air infiltration in an effort to increase the thermal integrity of a building, newer materials, such as heavier insulation (both batting and loose fill), have been introduced, as well as more advanced building wraps and vapor barriers.

Quantification of the integrity of the building envelope, achieved with the introduction of blower door testing, is another way to address the problem of heat loss. Blower door testing uses a mechanical fan to pressurize the building envelope and to also measure and quantify the amount of air leakage from the structure. By knowing the "before and after" amount of air leakage and using the blower door testing apparatuses to measure and compare this leakage, home energy contractors have been able to decrease the amount of air leakage from a structure by sealing such air leaks around windows, attics, and other building openings that in the past would have allowed air to infiltrate the building.

The goal of blower door testing and energy audit programs by state and federal government organizations, as well as private contractors, was and is to increase the long-term energy savings of the structure to reduce overall fuel consumption. An additional benefit is that it increases the property value of these homes.

Creation of a Negative Pressure Within the Building Envelope

By reducing the infiltration air openings into a building envelope, the building is able to develop and sustain a greater magnitude of negative air pressure in relation to the exterior of the home. Measurements of the differential pressure between the inside and outside of the home (Figure 8-19) show that there are periods of sustained negative pressure when the house is depressurizing and is essentially under a slight vacuum.

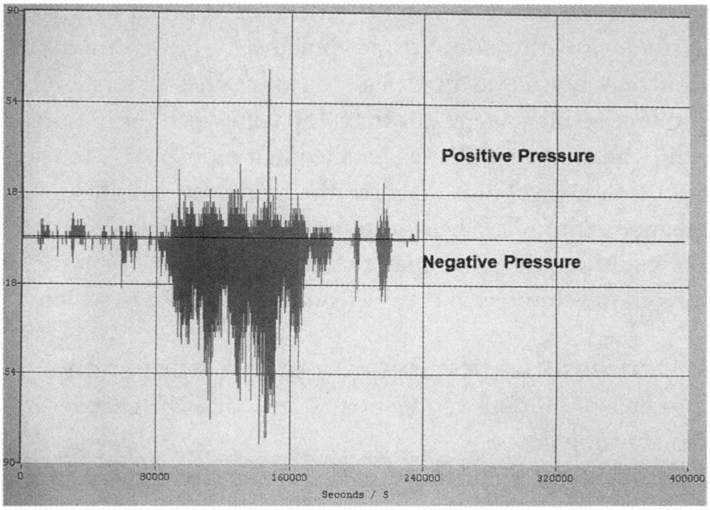

FIGURE 8-19 Differential Pressure Graph Shows the Development of Negative Pressure in Building Envelope (Courtesy of ISE Engineering Inc.)

Negative air pressure is developed in homes during winter months as a result of many factors. As an example, mechanical air handling devices such as fans and blowers create negative pressure by removing air from the interior of a building and depositing the air on the building's exterior. So do house fixtures and appliances such as bathroom fans, clothes dryers, and kitchen exhaust hoods while they are in operation. As well, combustion appliances such as boilers and furnaces, many of which are outfitted with induced draft fans, contribute to the building's ability to develop a large negative pressure in relation to the exterior of the home.

Another factor that contributes to the development of negative pressure in a house is wind flow over the structure. Airflow over a structure is unique since each house is surrounded by its own topography, which causes laminar and turbulent wind flow over the structure. These air currents have the ability to affect the internal pressure inside the structure.

Finally, air density differences between the exterior and interior of the home also affect the structure's ability to develop a negative pressure during winter months.

Combustion Appliance Issues and Depressurization

Changes in design and installation practices of combustion appliances using greater quantities of combustion air as well as the introduction of mechanical and direct venting, such as boilers and furnaces, are another contributor to depressurization-induced combustion reentry problems. These conditions make the system less resilient to other than the system running optimally since in the event of an improperly operating boiler or furnace, the force in which products of combustion will pass in and then out of a chimney, will decrease.

It is generally accepted that a depressurization of 0.020 inch of water or more can cause interference with natural draft appliances, and less harmful intermittent depressurization may be caused by exhaust fans, clothes dryers, and other exhaust devices that produce an airflow greater than 160 cubic feet per minute.

Additionally, an appliance that is installed in a room that is too small to provide adequate combustion air will also decrease the quality of combustion. This decrease allows the appliance to produce a heavier, more carbon-particle-laden product of combustion, which would require a comparatively stronger force to discharge the emissions not only from the chimney and away from the structure in which it is connected.

Backdraft—Depressurization Reentry at the Appliance

One path of product-of-combustion reentry is caused by the changes in airflow when the appliance is shutting down.

If a naturally vented combustion appliance is located in an area of depressurization, establishing or maintaining a proper draft is likely establishing or maintaining a proper draft is likely while it is firing and in operation. A firing appliance develops draft in the chimney, and a resulting higher force known as the stack effect is created. This stack effect will be great enough to overcome any building air pressure differences, allowing adequate operation of the heating appliance without combustion product spillage. This changes, however, when the appliance is shutting down.

During shutdown, there are still emissions in the chimney flue between the appliance and the peak of the chimney. Upon the termination of combustion, the negative air pressure effects within the structure will reverse the direction of flow of these emissions. This causes these products of combustion to back out of the chimney flue and appliance and into the home until the building air pressure differences are equalized.

This type of problem manifests itself as emissions that are seen backing out of heating system components such as barometric dampers and draft hoods as well as out of combustion chambers and fireplace dampers. See Figure 8-20. In gas-fired appliances, this is known as *rollout*.

It is not uncommon to see soot-stained insulation (Figure 8-21) in a basement where a depressurizing boiler or furnace is the source of combustion reentry. This is the result of continual cycles of reentering products of combustion discharging from any open breaching damper once the appliance shuts down following a heating cycle.

FIGURE 8-20 Measuring for Product of Combustion Flow Reversal

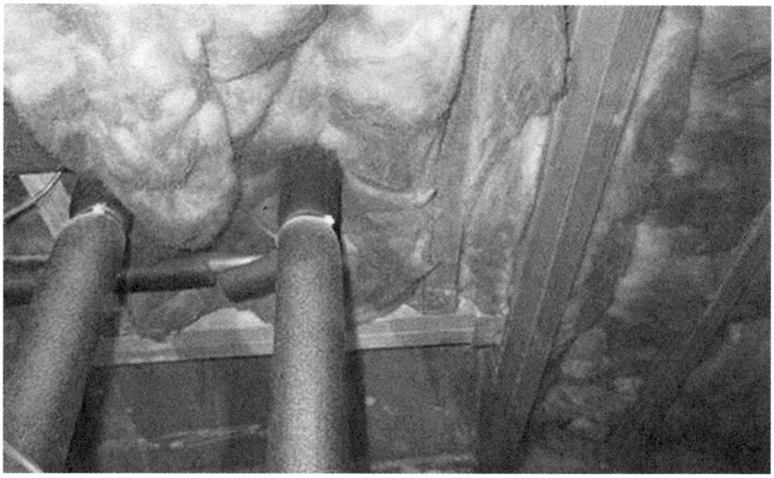

FIGURE 8-21 Soot Stained Insulation Above Boiler

Depressurization Reentry Due to the Proximity of Discharge

Depressurization reentry due to the proximity of discharge occurs when two conditions are met. One is that the building is developing a negative pressure. The other is that the products of combustion are discharging from a chimney or direct-vent appliance discharge hood that is exhausting products of combustion adjacent to the structure. These two conditions together create a situation where the products of combustion, although discharging from the system adequately, will be redrawn into the structure. Over time, the products of combustion will accumulate to the point where they become more noticeable.

Proximity of Discharge—The Chimney

The design of the chimney is crucial in thinking of the heating appliance and house structure as a "system," and many problems with product-of-combustion reentry into homes are the result of chimney construction and design.

Chimneys are designed to conform to building code or NFPA 211's guidelines for chimney construction, in that the peak of the chimney must be 2 feet over the top of the roof ridge where it is installed (when the chimney is within 10 feet of the peak). Frequently, though, the plume of emissions may come into contact with the building, which in turn allows negative air pressure to draw smaller amounts of these products of combustion back into the structure over time.

Another example of the proximity problem is a chimney that is properly built on a one-story section of the home (Figure 8-22). Oftentimes its discharge plume will strike a portion of the building that is higher than the one-story section. This allows products of combustion to come into contact with and then reenter the structure.

FIGURE 8-22 Chimney Installed on One-Story Portion of Building Allowing Discharge Plume to Srike Two-Story Portion of Building

Proximity of Discharge—The Direct-Venting Appliance

Another contributing factor to depressurization-induced combustion reentry problems is the use of direct-venting appliances (Figure 8-23). Direct-venting appliances became popular in the building boom of the 1980s because they provided significant cost savings in construction by eliminating the cost of chimney construction.

Unlike chimneys that discharge products of combustion into the airstream above the boundary layer of air around a structure, direct-vent appliances deposit these emissions adjacent to the building at or just above grade level in homes. Unfortunately, poor chimney placement and design or the unique features of the direct-vent appliance's installation, such as wind flow, negative pressure, and topography, can enhance the likelihood of reintroducing these substances into the structure.

FIGURE 8-23 Power Venter Direct Venting Mechanical Fan

Paths of Reentering Emissions Flow

In combustion reentry conditions where the emissions discharge from a direct vent or chimney and then reenter the structure, combustion reentry is evident first in the concentrations of soot-laden material in areas farther *away* from the appliance rather than at the appliance. Soot that has accumulated on insulation in an attic is a case in point.

An accumulation of soot on insulation materials in an attic is the result of emissions exiting the appliance's direct vent and reentering through roof openings and into the attic space (Figure 8-24). As negative air pressure draws these emission particles into the attic space, the particles are "filtered" through the insulation's fiberglass strands, leaving a blackened deposit on the insulation.

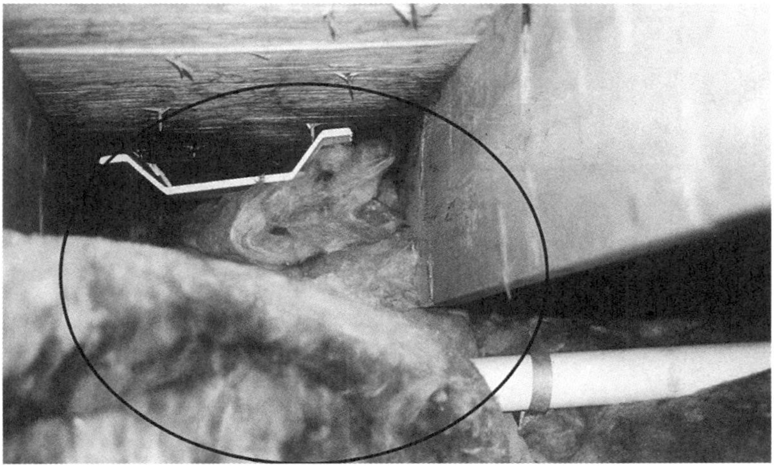

FIGURE 8-24 Soot-Stained Insulation in Attic

Negative pressure can also draw emission-laden reentry air from attic locations, through pathways created for house wiring, allowing emission particle reentry into the lower floors, as seen by the stained electrical outlet in Figure 8-25. Staining of carpeting adjacent to wall framing is also indicative of a leakage path through interior wall framing.

FIGURE 8-25 Soot Particles Emitting from Wall Box

Another common pathway of reentry emissions from an attic location is along vertical penetrations in the building envelope such as plumbing stack vent piping and attic hatch entryways (Figure 8-26).

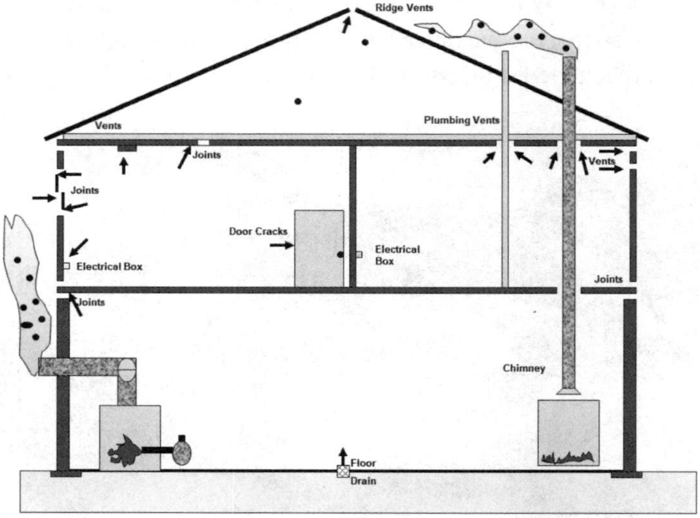

FIGURE 8-26 Depressurization Re-Entry Paths in Residential Structure

Recap of Symptoms of Depressurization-Induced Combustion Reentry Damage to Homes

Typically, in depressurization-induced combustion reentry conditions, only a small amount to no soot at all is discharged directly from the appliance, just as would be found in a conventional heating or combustion appliance malfunction.

Although small amounts of soot from oil-fired appliances may be found at places such as the barometric damper or other openings, usually homes with depressurization-induced combustion reentry show the characteristic heavier concentrations of soot-like substances *farther away* from the appliance and on higher floors rather than a direct discharge of soot into the home from the appliance itself.

Typically, homes that are being damaged from product-of-combustion reentry have concentrations of a soot-like substance in the attic, especially around vertical penetrations through the building envelope, such as vent pipes, as well as around holes drilled in the wall sill plates to accommodate wiring paths.

More extensive testing of the potential back-drafting of appliances has been suggested in documents such as the *ASTM E1998-11 Standard Guide for Assessing Depressurization-Induced Backdrafting and Spillage from Vented Combustion Appliances.*

Modifications to Reduce the Effects of Depressurization

The problem with depressurization-induced combustion reentry occurs as a result of two factors.

The first factor is the development of a negative air pressure condition in the building, coupled with the proximity of discharge of the appliance venting—whether it is a chimney or power venter termination point. There is also, however, the possibility that the chimney itself will backdraft, especially when the burner turns itself off at the end of a heating cycle.

Although each house is unique and a blanket statement cannot be made that will address all structures, in my experience, the most cost-effective as well as efficient way to address this problem is to extend chimney height (if so equipped) and make sure burners and equipment are properly maintained.

Increasing the chimney height will increase the stack effect of the chimney, which will help the appliance overcome any negative air pressure in the building. Additionally, a taller chimney will deposit the products of combustion up in the air-stream where they are more likely to be removed from the immediate vicinity of the building, and thus make it more likely that any infiltration air will be free of these substances.

The development of negative air pressure is more difficult to address since there are many sources of this condition, other than internally operating fans. With that said, different companies market heat recovery ventilators. These ventilators have fans that bring air from the outside into the building's structure when the combustion appliance is in operation in an effort to reduce the negative air pressure, thereby reducing or eliminating this problem.

References

ANSI Standard Z21.47: Standard Guide for Assessing Depressurization-Induced Back drafting and Spillage from Vented Combustion Appliances Gas Fired Central Furnace.

ASHRAE Handbook Fundamentals, 2013 edition.

ASTM D1356: Terminology Relating to Sampling and Analysis of Atmospheres.

ASTM E631: Terminology of Building Constructions.

ASTM E779: Test Method for Determining Air Leakage Rate by Fan Pressurization.

ASTM E1998-11: Standard Guide for Assessing Depressurization-Induced Backdrafting and Spillage from Vented Combustion Appliances.

Brook, David, "Putting Pressure on Building Codes," *Home Energy Magazine*, September/October 1996.

CAN/CGSB Standards, CANCGSB 51.71: The Spillage Test.

CMHC Laboratory Depressurization Test for Residential Gas Appliances.

CMHC Testing Oil-Fired Appliance Depressurization Spillage.

Godish, Thad, *Indoor Air Pollution Control*, Lewis Publishers, 1989.

NFPA 54, CAN\CG5B-51.74.

NFPA 211: Standard for Chimneys, Fireplaces, Vents, and Solid Fuel-Burning Appliances.

NFPA Technical Committee on Liquid Fuel Burning Equipment, meeting minutes, paragraph 31, June 25–26, 2009.

CHAPTER 9

HEATING SYSTEM CONTROLS AND REQUIRED PREVENTATIVE MAINTENANCE

A boiler and a furnace are both essentially heat exchangers. They differ only in the medium they are heating (water or air) and the energy source (gas, oil, electricity) they consume.

As stated earlier in the book, the heat exchanger and the combustion gas path have to be evaluated for leaks and cleanliness to make sure they are free of obstructions. Likewise the burner component has to be maintained to make sure it is burning fuel cleanly, efficiently, and safely.

This, however, is not all that has to be done in maintaining a heating system. In fact many problems and malfunctions do not originate from a heat exchanger or burner malfunction but rather are caused by a malfunction of the controls that govern their operational parameters such as water level, temperature, and pressure. Many controls are mechanical in nature, and as a result of issues such as maintenance neglect, age, or electrical supply, these controls may fail, which in turn can result in damage to the appliance or damage to the home in which the appliance is installed.

As a service technician, where do your responsibilities begin and end? Is a burner and boiler fireside cleaning all that is needed for you to meet your responsibilities in maintaining the appliance, or do your responsibilities extend to the auxiliary components that work in conjunction with the appliance?

What about maintaining the appliance's controls? Do you know that the homeowner may be tasked with the responsibility of maintaining some of the auxiliary safety controls? Do you know that the service technician is also required to perform maintenance on some of the auxiliary boiler and furnace controls?

Suppose you are maintaining a piece of equipment and the expectations of the property owner are such that your service work will keep the appliance and system operating effectively, efficiently, and safely. If a malfunction occurs due to a control or auxiliary component failure, who's at fault?

Needed maintenance actions aren't something that is specified by the homeowner. The homeowner is not the expert in maintaining the appliance whether he or she owns it or not. *You are!* Do not rely on a homeowner to tell you what has to be done. It's your job to tell the homeowner what you are required to do and explain why you are doing it and what can happen if you are not allowed to do your job thoroughly as intended by the manufacturer.

Remember, if something happens to the appliance, it may be unavoidable. However, it likely could have been prevented through proper cleaning and needed preventative repairs. And don't forget that in doing maintenance you need to check all components of the system, including the safety and auxiliary controls, and make sure you are knowledgeable of what is needed in both maintenance actions and spare parts to perform proper maintenance.

Control Parameter—Temperature

Low-Water Cutoff Switch

The job of a low-water cutoff switch is to protect a steam or forced hot water boiler from overheating. The low-water cutoff switch does this by verifying that water is present in the boiler. In the event that there is low or no water in the boiler, the cutoff switch stops the burner from firing. It may also simultaneously close a second circuit to a water admission valve control (known as a water feeding control), allowing the valve to add water to the boiler until the low-water cutoff switch senses this water and is satisfied.

The control is mounted at the boiler's lowest permissible waterline (Figure 9-1) and can be connected to the boiler by way of piping. It may be mounted externally or mounted directly to the boiler's pressure vessel.

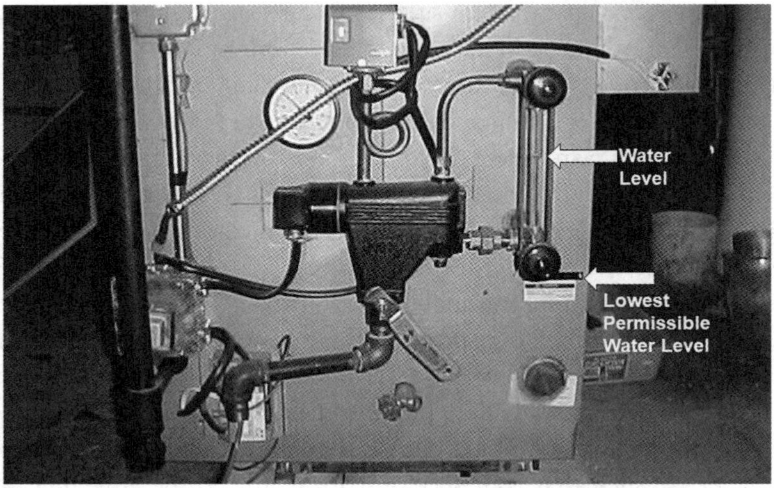

FIGURE 9-1 Float Type of Low-Water Cutoff Switch on Steam Boiler (Courtesy of McDonnell Miller–Xylem Inc.)

There are two main types of low-water cutoff switches: the float type and the probe type.

Float Type

The float type of low-water cutoff switch is just that, a control that operates using a float.

Description The low-water cutoff switch uses a float that is mechanically connected to an electrical switch connected to an oil or gas burner (Figure 9-2). If the water level falls, so does the float, which in turn opens the electrical circuit to the burner, stopping the burner from operating. Electrical connections (Figure 9-3) to the burner system and possibly the water feeder control allow these devices to coordinate the safe operation of the boiler and burner to prevent overheating.

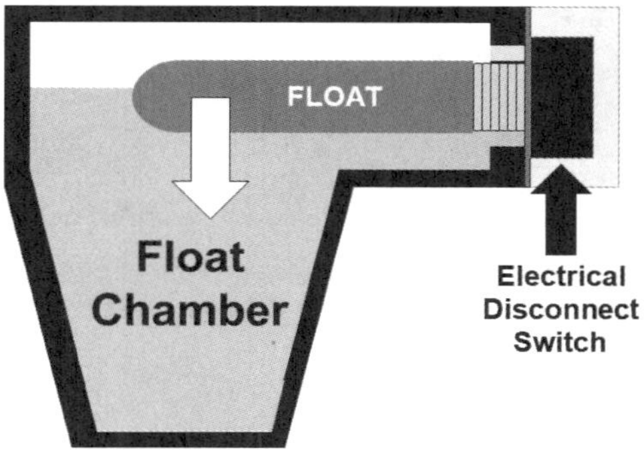

FIGURE 9-2 Illustrated Cross-Section of Float Type Low Water Cutoff Switch

FIGURE 9-3 Electrical Connections to Low-Water Cutoff Switch (Courtesy of McDonnell Miller–Xylem Inc.)

There are many designs and configurations of float type of low-water cut-off switches with design modifications for commercial or residential boilers. These designs may incorporate a larger float as well as an integrated water feeding valve (Figure 9-4).

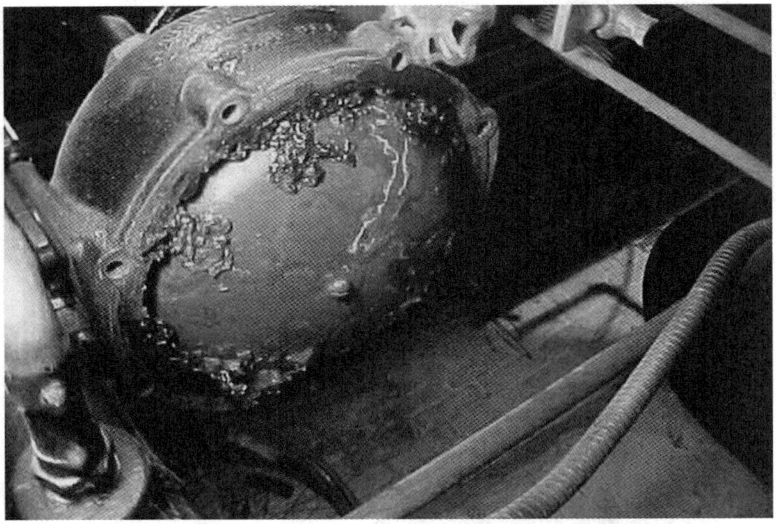

FIGURE 9-4 Integrated Float Type and Automatic Water Feeder Low-Water Cutoff Switch (Courtesy of McDonnell & Miller–Xylem Inc.)

Required Maintenance Maintenance should be done according to Section VI of the *ASME Boiler & Pressure Vessel Code*: "Low water fuel cutoff and water feeders should be dismantled annually, by qualified personnel, to the extent necessary to insure freedom from obstructions and proper functioning of the working parts" (paragraph 7.07 G).

Leaks Water leaking out of the low-water cutoff switch will cause the electrical disconnect switch to seize and will produce stray currents that can cause the control to remain closed in a low-water situation. To prevent this problem, during annual or preventative maintenance, examine any gasket areas sealing the water-bearing float chamber area to identify any leaks and repair them (Figure 9-5).

Any water contact with crucial operational parts such as the control's micro switch (Figure 9-6) should prompt the replacement of that component or the entire control.

Sludge and Scale As boiler water turns to steam, which in turn travels through the system, leaves only if the system is leaking. The hardness minerals in the water will concentrate and turn from a liquid form into a solid. These solids, which are now mixed with only a very small volume of water, form a mud-like slurry called sludge that will settle in the lower portions of a steam boiler and in the float and float chamber of the low-water cutoff switch (Figure 9-7).

FIGURE 9-5 Water Leaked into This Control's Electrical Switch Due to Lack of Preventative Maintenance (Courtesy of McDonnell & Miller–Xylem Inc.)

FIGURE 9-6 Water Damage to Control's Micro Switch (Courtesy of McDonnell & Miller–Xylem Inc.)

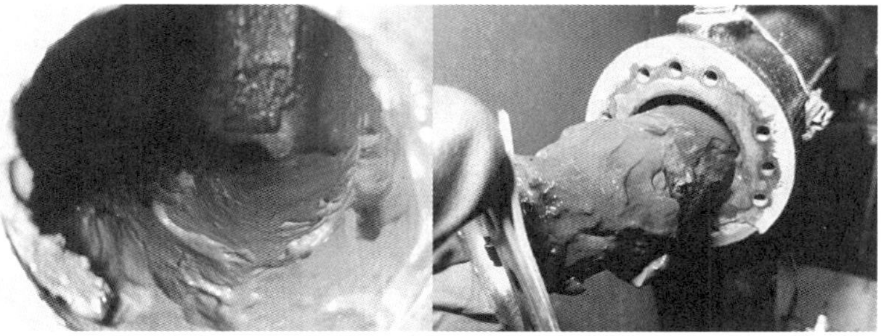

FIGURE 9-7 Sludge and Scale Build Up in Low-Water Cutoff Switch Float Chamber and on Float (Courtesy of McDonnell & Miller–Xylem Inc.)

During the heating season, this sludge must be removed on a weekly basis by the homeowner. The homeowner's job is to open the flush valve on the control, which in turn drains a small amount of boiler water along with sludge from the float chamber (Figure 9-8). Since the sludge has a higher water volume than the solid, it will freely flow with the drained boiler water during these flushing events.

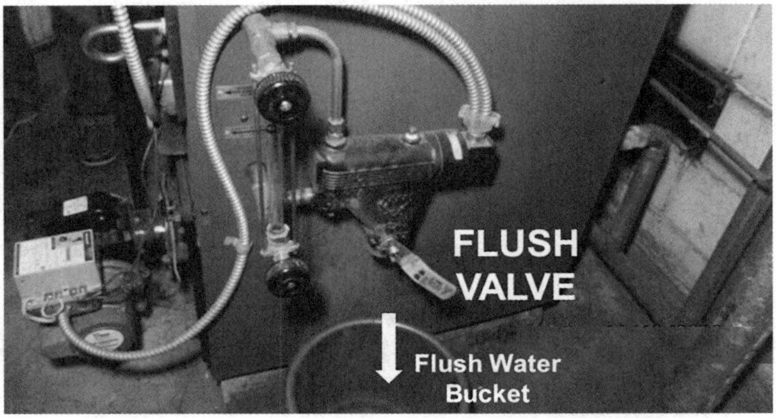

FIGURE 9-8 Flushing Low-Water Cutoff Switch
(Courtesy of McDonnell & Miller–Xylem Inc.)

Manufacturers will often provide instructions to their customers. For example, McDonnell & Miller–Resideo attach a "Notice" tag directing the weekly flushing of low-water cutoff switches (Figure 9-9). The tag instructs the homeowner to "Place bucket under valve and blow down this control at least once each week during heating season." It then goes on to warn the homeowner that failure to follow the blow-down procedure on boiler card furnished can cause control to malfunction, resulting in serious boiler damage.

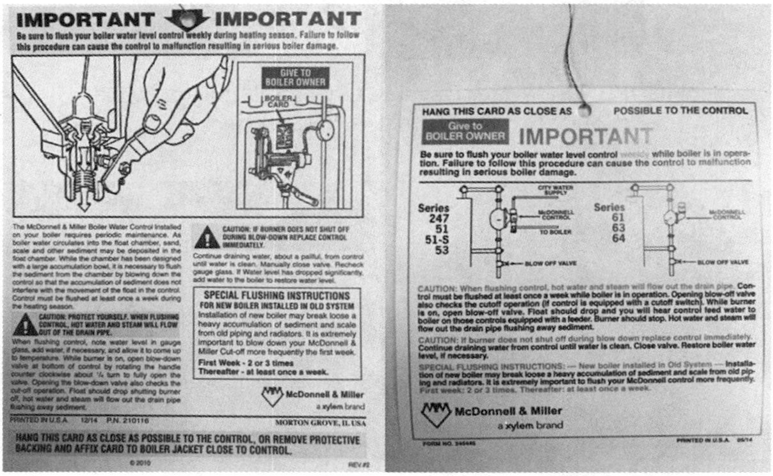

FIGURE 9-9 Low-Water Cutoff Switch Maintenance Instructions
(Courtesy of McDonnell & Miller–Xylem Inc.)

If the sludge is not reduced by the weekly flushing of the low-water cutout switch or if there are areas of the control where the flushing action does not remove all the sludge impurities, over time the sludge will dry out and form a hard scale (Figure 9-10). This is a more difficult substance to remove than the sludge. And once there is a large accumulation of scale, it cannot be dislodged and flushed from the control by the flushing action alone (Figure 9-11). The consequence is that the float in the low-water cutout switch will be impeded from traveling in the vertical direction in a manner that coincides with the true boiler water level.

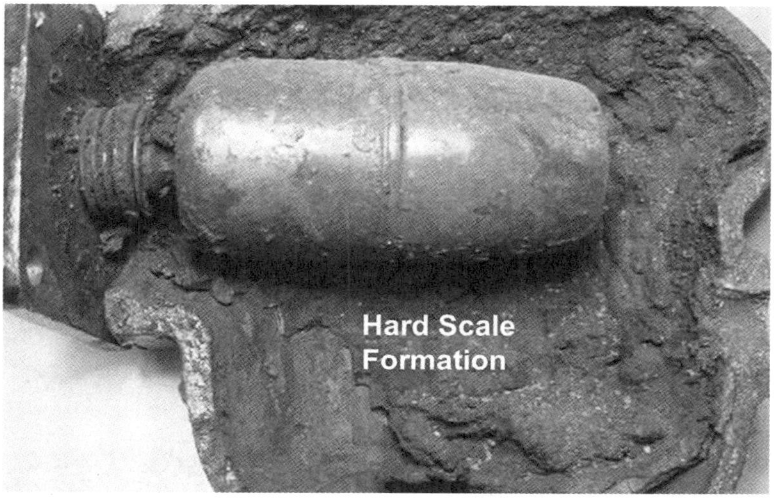

FIGURE 9-10 Bisected Low-Water Cutoff Switch (Courtesy of Watts Industries)

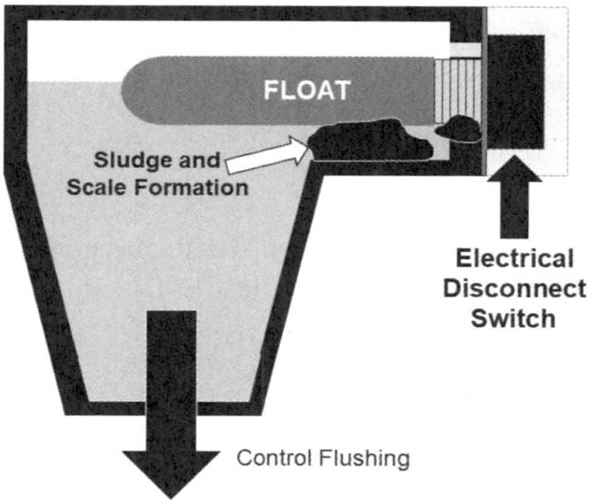

FIGURE 9-11 Sludge and Scale Formation in Float Type of Low-Water Cutoff Switch

Disassembly Remember the earlier quotation from Section VI of the ASME Boiler & Pressure Vessel Code that begins "Low water fuel cutoff and water feeders should be dismantled annually"?

"Dismantled annually." Get it? You have to take the equipment apart. Disassembly and proper cleaning on an annual basis, as directed by the manufacturer and standards such as Section VI of the *ASME Boiler & Pressure Vessel Code*, will remove the hard scale accumulations that normal, weekly flushing of the device cannot remove.

As shown in Figure 9-12, controls are designed and supported by the manufacturer to be maintained in an efficient and safe manner. Spare components such as gaskets should be on hand, allowing quick turnaround of repairs.

FIGURE 9-12 Disassembly and Cleaning of Low-Water Cutoff Switch (Courtesy of McDonnell & Miller–Xylem Inc.)

Probe Type

The other type of low-water cutoff switch is the probe type of control. Rather than using a float to verify that there is water in the boiler, the probe type of low-water cutoff switch uses a water sensor (Figure 9-13).

Description The probe type of control is connected to the boiler's pressure vessel at the waterline, and the control's electronics use the signal provided by the sensor to turn the burner off or on, depending on the water level, as well as to activate a water feeder signal.

The probe type of low-water cutoff switch is usually mounted in its entirety at the boiler's minimal allowable water level (Figure 9-14).

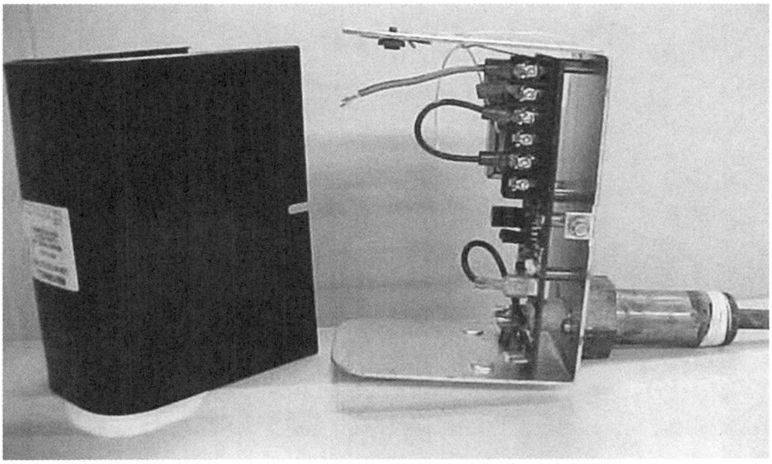

FIGURE 9-13 Probe Low-Water Cutoff Switch
(Courtesy of McDonnell & Miller–Xylem Inc.)

FIGURE 9-14 Probe Type of Low-Water Cutoff Switch Mounted on Steam Boiler

The sensing probe that is fastened to the boiler pressure vessel (Figure 9-15) has two conductive circuit paths that, when immersed in boiler water, complete the circuit "telling" the control that there is an adequate water level in the boiler. This allows the burner to fire in response to a call for heat.

FIGURE 9-15 Low-Water Cutoff Switch Sensing Probe

Maintenance Annually or as specified by the manufacturer, the probe of this control is required to be removed and cleaned to remove any accumulated sludge or scale buildup (Figure 9-16), which can interfere with the control's operation and allow a false water presence signal to occur. This is an easy task to complete and only requires the removal of the control electronics housing body and the probe.

FIGURE 9-16 Sludge and Scale on Low-Water Cutoff Probe

The probe can be cleaned with a soft cloth. At this time, careful inspection of the probe for any signs of material damage can be performed as well.

FORENSIC INSIGHT

Did This Thing Overheat?

Consequences of Control Failure of Low-Water Cutoff Switches

Whether the type of low-water cutoff switch is the float or the probe type, the consequences of failure are the same. An unmonitored boiler that is empty of water and is firing in response to a call for heat will cause catastrophic damage certainly to the appliance but often to the home it is installed in as well.

Signs of damage caused by overheating can frequently be seen on the outside of the appliance. For example, as Figure 9-17 shows, overheating of a cast-iron boiler is outwardly evident by the scorch marks on the sheet metal appliance cover.

FIGURE 9-17 Jacket Scorching on Overheated Boiler

Inside the boiler, overheating damage to the boiler castings will be evident by a combination of red and black-blue discoloration produced by heat-induced oxides of the iron in the boiler castings or heat exchanger. These heat-induced oxides (ferric and ferrous oxide) are indications that the castings got to a temperature of at least 1600°F, well beyond the design limits of the boiler.

Prolonged overheating of the castings will damage the boiler beyond repair by causing cracks to form (Figure 9-18). In boilers where gaskets seal the sections, prolonged overheating of the castings will cause the gaskets to melt.

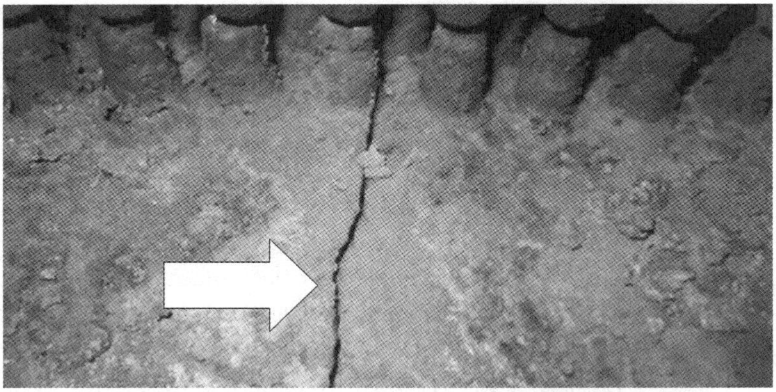

FIGURE 9-18 Overheating Caused Boiler Section to Crack

It doesn't stop there. Steam boilers are found in homes with steam heating systems that haven't been popular since the early part of the twentieth century. The homes they are installed in are usually older, and many have lumber with a rough-sawn surface. Rough-sawn lumber, with its protrusions of wood fibers, burns more easily than modern, smooth-surfaced lumber.

Couple this fact with the hazard posed by an empty, malfunctioning boiler. The burner is constantly firing because the thermostat is calling for heat, but the heat will never reach the other areas of the home because the boiler has no water in it to make steam to carry the heat there. What you get is . . . a very dangerous situation!

An overheating boiler that is dry-firing due to a failed low-water cutoff switch can and will cause fires (Figure 9-19), especially if the boiler's clearances to combustibles is not installed in accordance with local codes and the manufacturer's instructions.

FIGURE 9-19 Low-Water–Fired Boiler Caused Adjacent Combustibles to Ignite

Aquastat (Forced Hot Water Boilers)

Different from how a steam boiler prevents overheating, a forced hot water boiler maintains the temperature of the water in its system with a temperature-regulating control known as an aquastat. (Note that some forced hot water boilers are required to have probe-type low-water cutoff switches rather than aquastats, depending on local jurisdiction code requirements.)

Description Initially, aquastat design was of the capillary tube type of control (Figure 9-20). In this type of control, the thermal sensing bulb and attached liquid-filled capillary tubing responded to changes in temperature, which in turn activated switches in the control.

FIGURE 9-20 Capillary Tube Type Aquastat (Courtesy of Honeywell–Resideo)

Depending on the control model, the design of the aquastat has advanced so that the temperature adjustment, circulator, burner, and even primary control safety functions can be accommodated in one integrated control as opposed to several. And more recent versions of the aquastat incorporate electronic (noncapillary tube) technology as well as LED display diagnostics to assist in troubleshooting equipment failures (Figures 9-21 and 9-22).

The sensing element or bulb of the aquastat is attached to the boiler at an immersion well that is connected to the boiler by a threaded connection (Figure 9-23). A sensing paste may be applied to the sensing element in the aquastat well to enhance thermal transfer. The aquastat immersion well is particular to each aquastat control, and not all immersion wells are compatible with replacement controls.

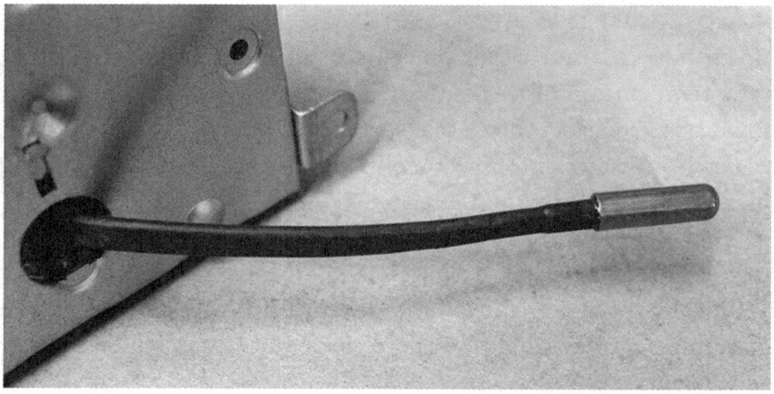

FIGURE 9-21 Electronic Aquastat (Courtesy of Honeywell–Resideo)

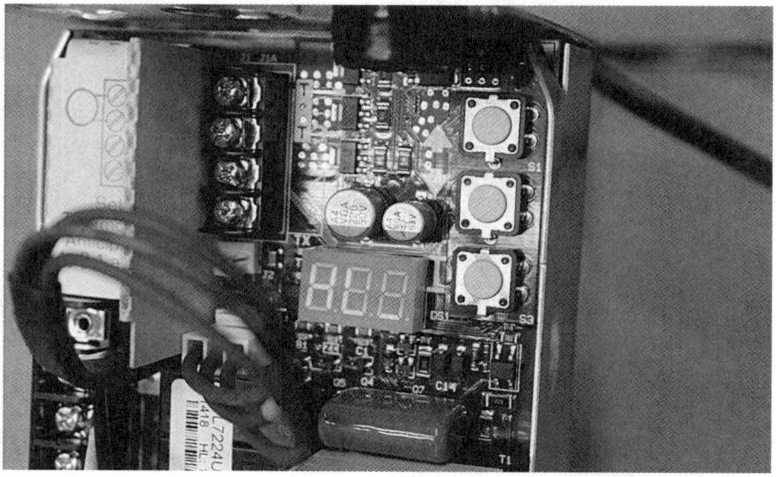

FIGURE 9-22 Aquastat LED Display (Courtesy of Honeywell–Resideo)

FIGURE 9-23 Immersion Well for Aquastat (Courtesy of Honeywell–Resideo)

As seen in Figure 9-24, the well is immersed in boiler water in the boiler section casting or heat exchanger. This assures uniform heat transfer to the sensing well, and it also assures representative temperature sensing of the aquastat.

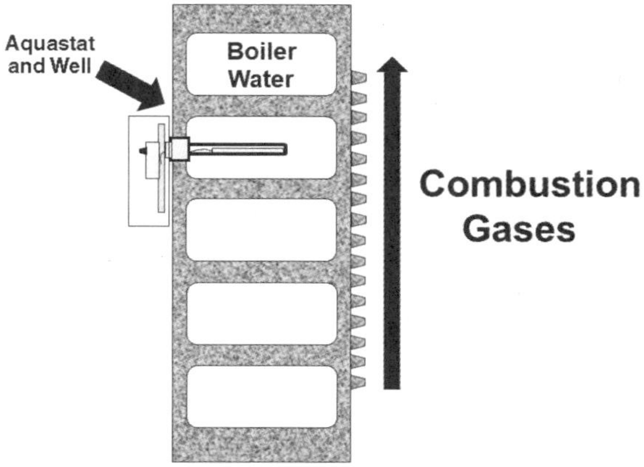

FIGURE 9-24 Positioning of Aquastat Well in Boiler Section

The capillary bulb or element is inserted into the well, and the control body is secured to the end of the well (Figure 9-25).

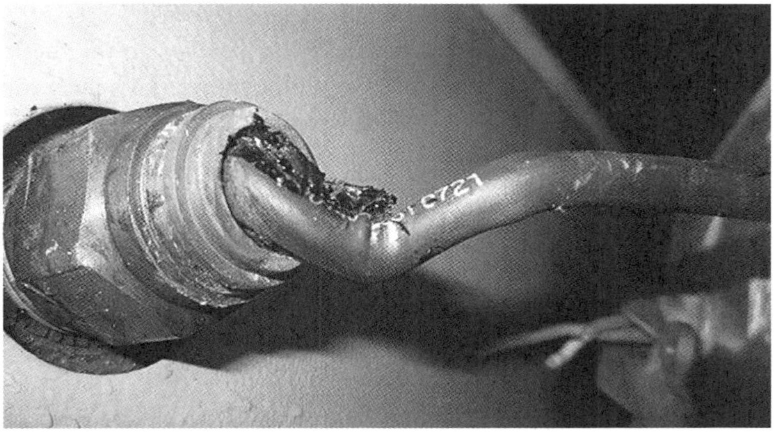

FIGURE 9-25 Heat Sensing Element Inserted into Aquastat Well (Courtesy of Honeywell–Resideo)

Maintenance The performance of the control can be verified during routine maintenance with the use of a thermocouple, infrared, or noncontact temperature-sensing device to determine the operating temperature of the boiler. By monitoring the boiler's supply and return temperatures against the high and low aquastat limits through at least one complete cycle, the technician can verify that the controller operates as intended. If there is a discrepancy, the aquastat settings should be examined and the control replaced if needed.

Furnace High-Limit Switches (Forced Hot Air Furnaces)

The exit temperatures of the air from a forced hot air furnace are monitored and controlled by a high-limit switch.

Description The high-limit switch is placed downstream of the air supply from the furnace and controls the burner as well as the blower functions, depending on the furnace design (Figure 9-26).

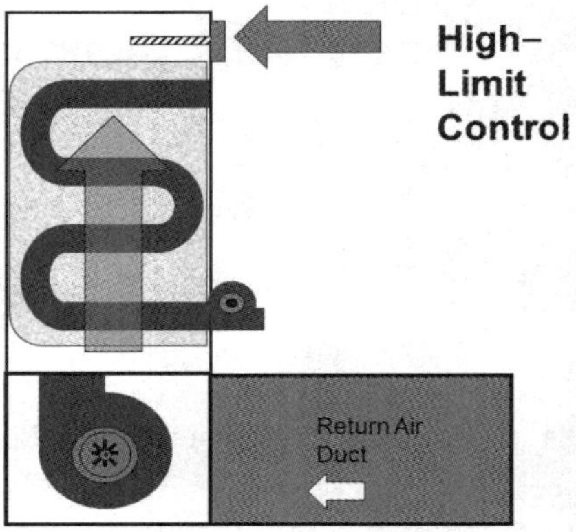

FIGURE 9-26 Positioning of High-Limit Switch in Forced Hot Air System

There are essentially two types of control designs.

The first is the helical coil mechanical type, which relies on the thermal expansion of a metal coil that responds to changes in temperature. The helical coil, which is inserted into the airstream, senses the different temperatures of the exiting airstream. These different temperatures correspond to various control settings coordinating the furnace's operation.

Typically, the high-limit setting, which causes the burner to stop operating, is in the 200°F range (Figure 9-27).

FIGURE 9-27 Helical Coil Type of High-Limit Control (Courtesy of Honeywell–Resideo)

Variations of this control include fan power switches that allow the blower fan to be turned on regardless of the thermostat or control setting as well as switches that have adjustable temperature settings for high temperature limits and a blower fan operating temperature range (Figure 9-28).

FIGURE 9-28 Temperature Settings on Fan-Limit Control (Courtesy of Honeywell–Resideo)

The second type of design of a furnace limit switch is the thermal disk type (Figure 9-29). This type of control positions the thermal-sensing portion of the control in the furnace airstream, exposing it to the heated air being produced by the furnace.

FIGURE 9-29 Thermal Disk Type Limit Switch

"Thermal switch" types of high-limit controls have no adjustable settings.

Maintenance There is no "maintenance" of these controls. However, verification of their correct operation is a good preventative maintenance decision since in the event of a limit control failure, damage to the furnace as well as a potential fire can ensue.

Likewise, in the same way an aquastat's performance is verified, the performance of the control can be verified by monitoring the air supply temperature with the use of a thermocouple probe inserted into the airstream in the plenum or furnace outlet and comparing the temperature with the control's dial or shutoff set point.

On some controls the blower power leads can be disconnected, allowing the control to be tested in a static, nonflowing airstream.

Consequences of Failure In the event of a high-limit failure, the furnace will produce excessive temperatures. This high heat may damage the furnace's heat exchanger, resulting in excessive expansion and crack development.

Excessive temperatures can also cause the furnace insulation to overheat and debris in the furnace and on the furnace's heat exchanger to ignite.

For furnaces that double as air handler fans for air conditioning, the evaporator coil and condensate pan can also overheat, causing soot discharge into the system ductwork as well as fires (Figure 9-30).

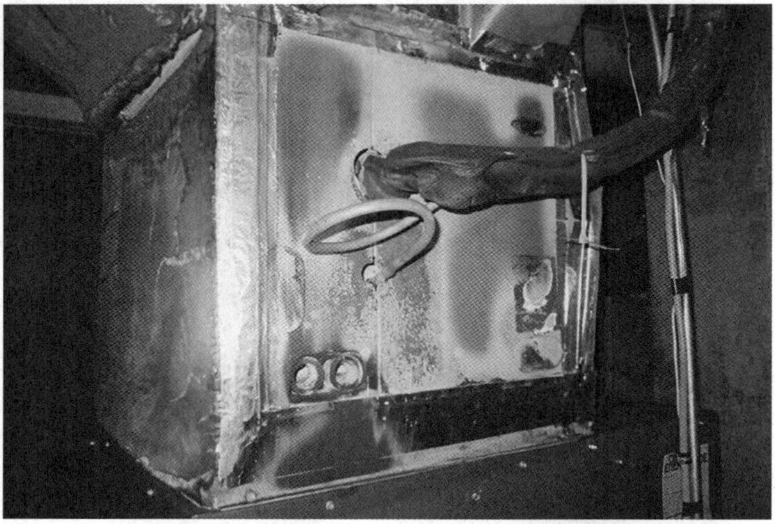

FIGURE 9- 30 Fire Damage Furnace Ductwork Air Conditioning Equipment Due to a High-Limit Failure

Control Parameter—Pressure

Pressure Control (Steam Boilers)

Residential steam systems operate at a pressure of 0.2 to 0.5 psi, and pressures are maintained with a pressure-regulating control (Figure 9-31).

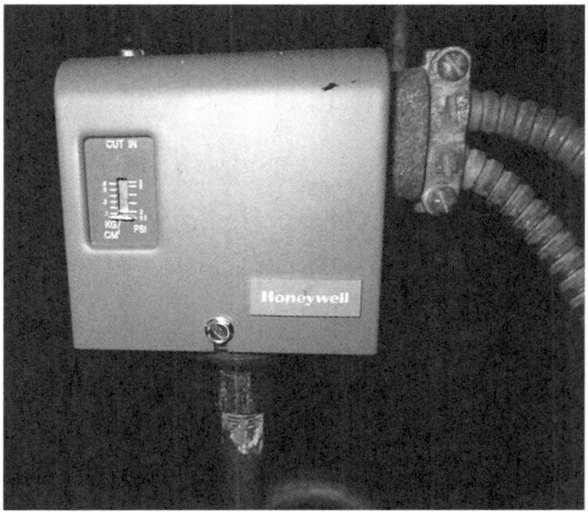

FIGURE 9-31 Pressure Control for Steam Boiler (Courtesy of Honeywell–Resideo)

Description The pressure control in a residential steam boiler is connected to the steam side of the boiler pressure vessel with a "pigtail" steam trap siphon loop (Figure 9-32) to protect the control from carryover elements from the boiler.

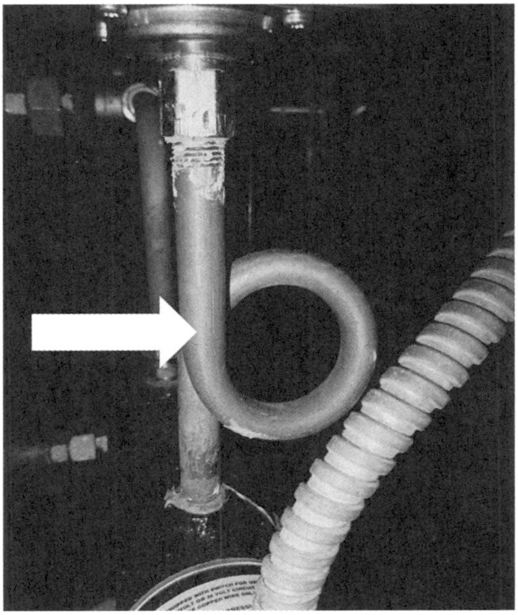

FIGURE 9-32 Pressure Control Siphon Loop

Maintenance The control can be visually inspected for proper operation during preventative maintenance events by raising and lowering its set point. Pressure should increase when the set point is raised and decrease when the set point is lowered.

Consequences of Failure In the event of a control failure or blockage in the pigtail steam trap, excessive pressure in the boiler may develop. Signs of overpressurizaton of the boiler include knocking of the steam pipes as well as relief of pressure from the boiler's safety relief valve.

Expansion Tanks (Forced Hot Water Boilers and Water Heaters)

An expansion tank allows a reservoir of compressed air into which system water can expand to reduce system pressure fluctuations.

Description An expansion tank is connected to the forced hot water system piping.

An older type of expansion tank is filled with air and has system water flowing into it, allowing it to compress this air. These older tanks were installed frequently throughout the 1950s and 1960s and are usually cylindrical in shape (Figure 9-33).

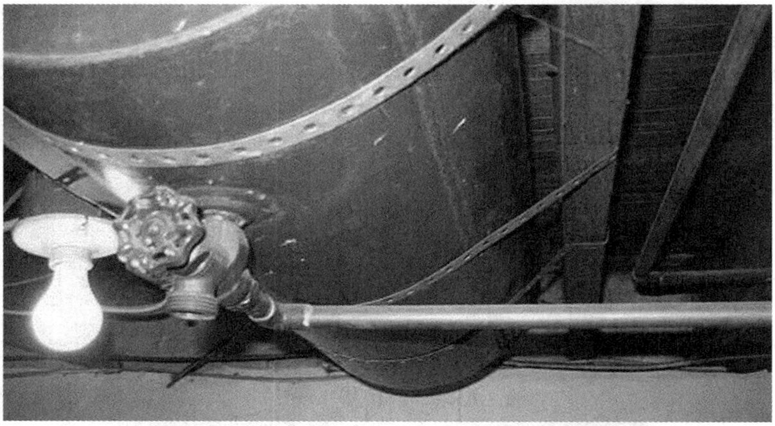

FIGURE 9-33 Drum Type of Expansion Tank

A modern and more common expansion tank is the diaphragm expansion tank. This type of tank is divided in half by a rubber diaphragm. One side of the tank is filled with the system's water, and the other side is charged with air. As the water in the system expands, the gas on the opposite side of this diaphragm is compressed (Figure 9-34).

Due to age, improper sizing, and water chemistry, diaphrap expansion tanks can flood completely with water. This prevents any expansion of water and dampening of system pressure, allowing pressures to rise above intended system levels.

Maintenance Diaphragm types of expansion tanks should have the air valve tested for pressure to make sure there is a proper charge present annually. This can be done with a tire pressure gauge (Figure 9-35). The air valve portion of the tank should not contain water, which will flow out of the valve if the valve is depressed. If water is present in this portion of the tank, the tank should be replaced. Ideally, the air pressure should be tested in diaphragm type expansion tanks by isolating the tank from the boiler if a valve between them is installed. The target pressure is to have the pressure of the air side of the expansion tank equal the pressure of the boiler waters operating pressure.

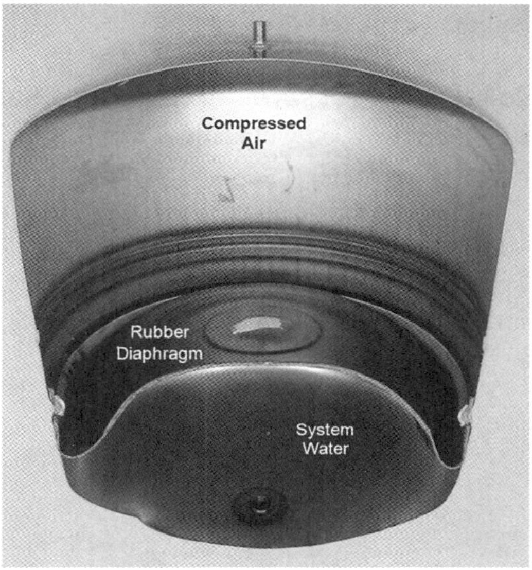

FIGURE 9-34 Bisected Diaphragm Type of Expansion Tank

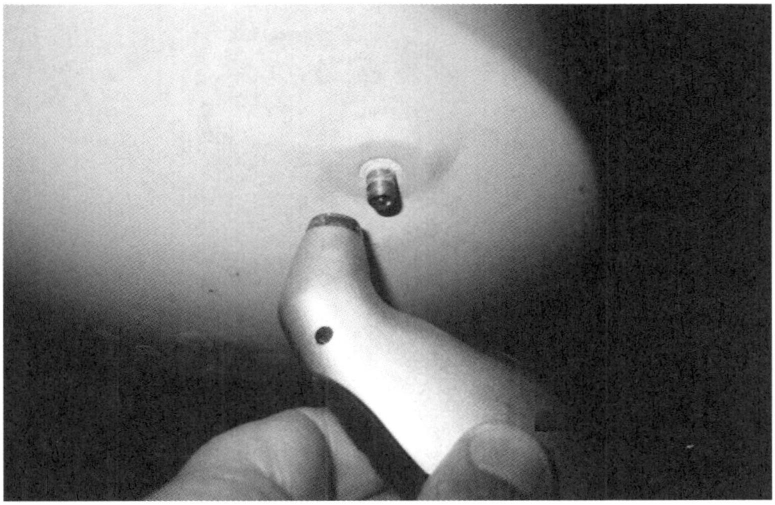

FIGURE 9-35 Checking Air Charge on Diaphragm Type of Expansion Tank

Older cylindrical expansion tanks should not be completely full of water and should be slightly drained if these are overfilled. System water must have an "air space" area within these tanks so that this space can receive any volume of expanding water.

Consequences of Failure Without a reservoir of expansion space for pressure fluctuations in the system, a boiler with a failed expansion tank will be reliant upon the boiler's safety relief valve to relieve internal pressures in excess of the system operating pressure. Indications of a failed expansion tank will be evident by dripping safety relief valves and steam leakage in the vicinity of the boiler. If gone unnoticed, this can lead to mold growth and building damage.

Safety Pressure Relief and Pressure-Temperature Safety Relief Valves

Two common types of relief valves you will come across are the boiler pressure safety relief valve and the domestic hot water temperature-pressure safety relief valve.

Boiler Pressure Safety Relief Valve

A boiler's safety relief valve is simply a spring-loaded, normally closed valve that opens when the vessel pressure overcomes the force of the valve spring (Figure 9-36). Upon the reduction of vessel pressure, the valve closes.

FIGURE 9-36 Bisected Safety Relief Valve (Courtesy of Watts Regulator Company)

Depending on the operating pressure of the boiler that the valve is protecting, the valve operates in a number of different pressure ranges. Typically, residential boiler pressure relief valves discharge at the 30-psi level.

Small fluctuations in system operating pressure due to heating and cooling of water during normal operating cycles is compensated for by an expansion tank. However, if the expansion tank fails or if the capacity of the tank is exceeded, the boiler is reliant upon the pressure safety relief valve to relieve excessive pressure and prevent the pressure vessel from being damaged.

Description A boiler safety relief valve is mounted directly onto the boiler pressure vessel with no valves between it and the boiler (Figure 9-37). Discharge from the valve is directed away from the appliance and into a suitable location through discharge piping.

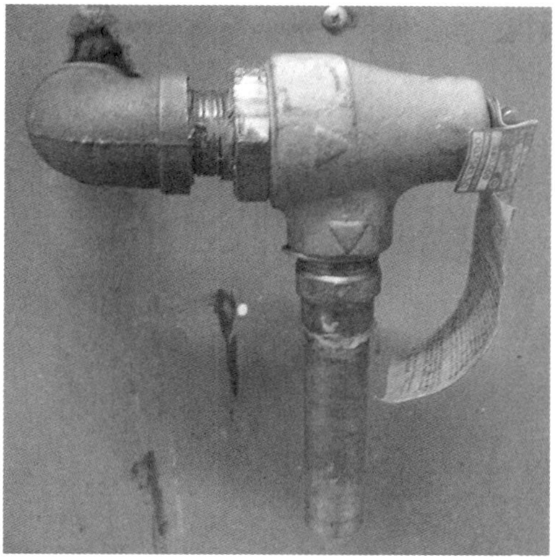

FIGURE 9-37 Discharge Piping Connected to Safety Relief Valve Diverts Discharging Water Away from Appliance (Courtesy of Watts Regulator Company)

The piping should never be obstructed or reduced in size (Figure 9-38). since this may affect the ability of the valve to relieve pressure as needed.

FIGURE 9-38 Discharge Piping Should Never Be Restricted in Size

Domestic Hot Water Temperature-Pressure Safety Relief Valve
For domestic hot water heaters that operate at a higher pressure and in which scalding and excessive temperature are a concern, temperature-pressure relief valves are installed. This type of valve is rated at 150 psi and is almost identical in design to a boiler's safety relief valve. The main difference is that it is outfitted with a stronger spring that allows it to lift at a higher pressure.

The pressure-temperature relief valve is also equipped with an "extension thermostat." This component is a wax-filled cylinder that has a spring-loaded pin (Figure 9-39). In the event of a temperature increase beyond 210°F, the pin will force the valve disk in an upward direction and cause the valve to open.

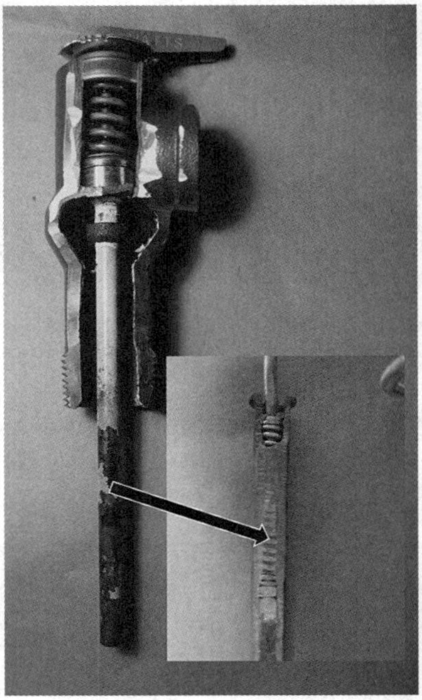

FIGURE 9-39 Temperature-Pressure Relief Valve (Courtesy of Watts Regulator Company)

Maintenance of Pressure and Temperature-Pressure Relief Valves Pressure-temperature safety relief valves let water escape if the pressures in boilers and water heaters exceed the valves' rated operating pressure. These valves occasionally open to allow water to exit the system. This usually occurs after the boiler has been turned on at the beginning of a heating season or after a mechanical breakdown.

Although many service technicians do not like to perform tests on the safety relief valve because they are afraid that the valve will not reseat properly, this maintenance must be done on a preventative basis to make sure the appliance is safely protected.

The relief valve should be checked for any evidence of preexisting and prolonged leakage, such as corrosion on the valve or discharge piping (Figure 9-40). If the valve shows evidence of leakage, flushing the valve and connecting piping with water should remove any impurities. If dripping continues, the valve should be replaced.

The same corrosion process or accumulation of impurities in the valve's inlet or outlet piping can also cause the valve to seize or become totally blocked. A seized valve can be identified and likely corrected by performing a "lever or pop test" by lifting the valve handle to make sure it is free of obstructions and will move freely and allow water to flow (Figure 9-41).

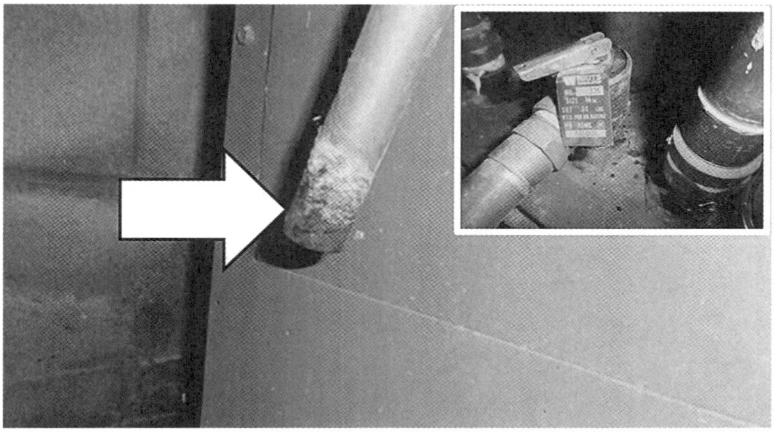

FIGURE 9-40 Corrosion at End of Discharge Piping May Be an Indication of a Constantly Leaking Safety Relief Valve (Courtesy of Watts Regulator Company)

FIGURE 9-41 Performing Lever Test on Safety Relief Valve (Courtesy of Watts Regulator Company)

In performing maintenance, be sure to check and follow pertinent codes and the manufacturer's instructions:

- **Code direction.** Although code direction usually addresses installation issues, there are several code directives that address the requirement to properly maintain heating appliances and auxiliary components. For example, Chapters V and VI of the *ASME Boiler & Pressure Vessel Code* call for annual maintenance and the performance of a "pop test" of the pressure relief valve to ensure that it is operating properly. And Chapter 1 of the *International Mechanical Code* provides the blanket statement that "Devices or safeguards which are required by this code shall be maintained in compliance with the code edition under which they were installed," paragraph 102.3.
- **Manufacturer's instructions.** Manufacturer's instructions for these components include maintenance advisories such as the one shown in Figure 9-42.

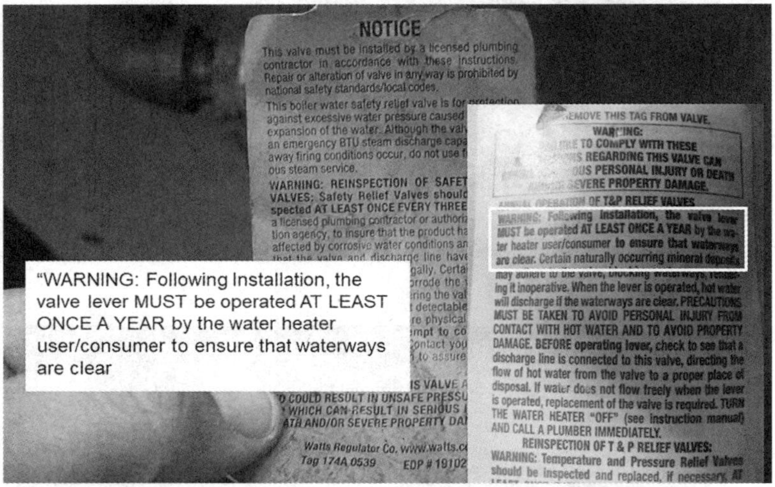

FIGURE 9-42 . Provided Maintenance Instructions (Courtesy of Watts Regulator Company)

Consequences of Failure Pressure vessel explosions often occur following a service event in response to a no-heat call. The inherent cooling and contraction of the water in a boiler or water heater during this downtime can set the stage for a "time bomb" to develop.

During downtime the system water cools and reduces the system pressure. Upon repair of the burner or appliance heat source, this water will then rise in temperature and expand. As the system's temperature and pressure continue to rise even further, the system now becomes reliant upon the operation of the safety relief valve.

If the safety relief valve is incapable of lifting, pressures approaching hundreds of pounds per square inch may develop, causing pressure vessel failure and the immediate vaporizing of the compressed and heated water within.

The effects of a burst pressure vessel can be as mild as a small split in a cast-iron boiler section or a displaced gasket. But the damage can also be very severe to both the boiler and the building in which it is installed (Figure 9-43). Examples include:

- Structural building damage
- Displacement of heavy cast-iron boiler
- Broken gas lines and fire
- Metal shrapnel projection
- Burns due to scalding water
- Injury and death

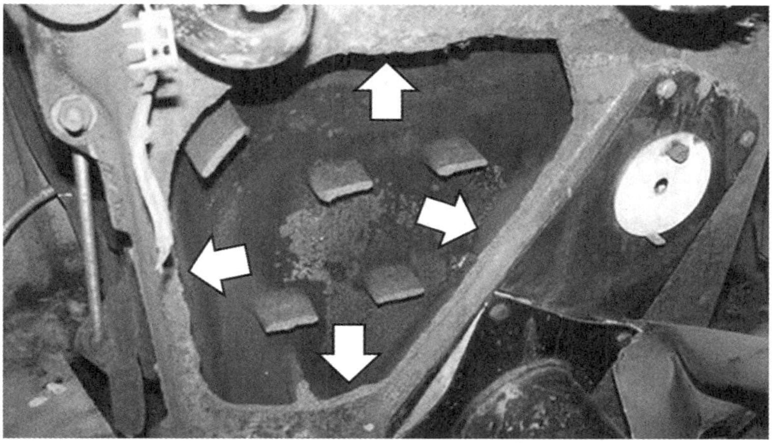

FIGURE 9-43 Boiler Damage Due to Over Pressurization

Control Parameter—Water and Steam Management

Forced Hot Water Boiler—Pressure-Reducing Water Feeder Valves

Forced hot water boilers operate at between 12 and 30 psi. Although, technically, forced hot water systems are "closed" in that they are not supposed to lose water, they do.

In many cases, water loss is caused by leaks in the expansion tank, leaking boiler sections, or a water discharge from the safety relief valve. To compensate for the loss of water, it is necessary to provide a constant source of makeup water to the boiler system.

Domestic water supplied to a building's plumbing system is typically between 65 and 80 psi. However, the pressure ceiling of a heating boiler is limited to 30 psi, and as such, the pressure of the supplied water has to be reduced to prevent nonsanitary boiler water from finding its way back into the domestic water source.

Description Water is supplied to forced hot water boilers through a backflow preventer and boiler feed water pressure reducing regulator where the pressure is reduced to boiler operating pressures (Figure 9-44). The pressure-reducing valve shown in

Figure 9-45 is outfitted with a pressure adjustment screw that applies force on top of the pressure-sensing diaphragm. The valve also has a manual hand-fill lever to allow faster filling of the system during maintenance and installation when the system has been drained. In the event that the boiler operation causes the pressure in the system to increase, water is prevented from flowing back into the domestic water system by two different valves. One is a check valve (Figure 9-46) that is built internally in the pressure-reducing valve, and the other is a backflow preventer valve upstream.

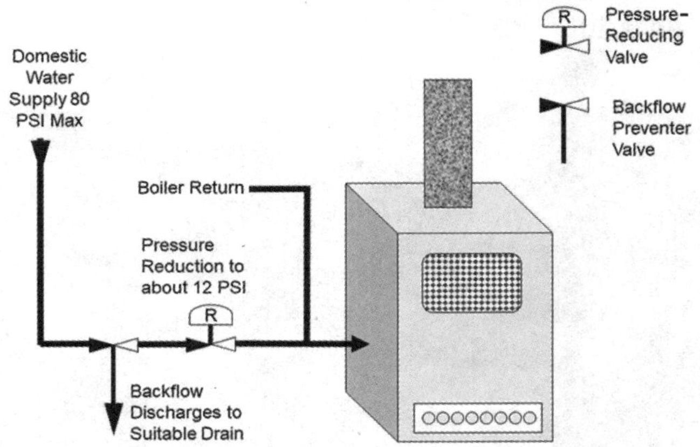

FIGURE 9-44 Makeup Water Addition to Forced Hot Water System

FIGURE 9-45 Pressure Adjustment and Fast Fill Lever (Courtesy of Watts Regulator Company)

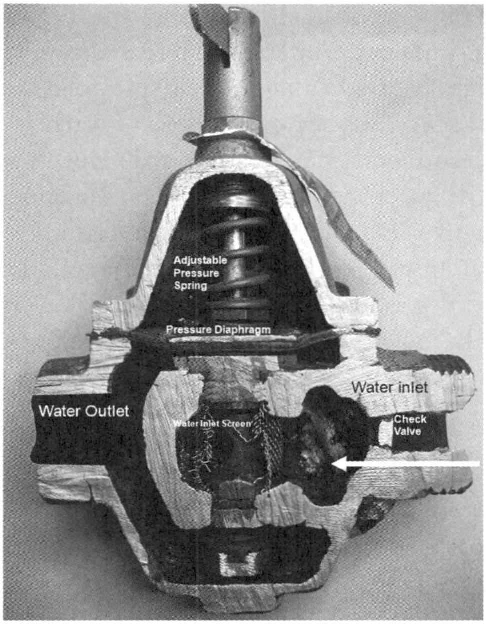

FIGURE 9-46 Cutaway View of Water Boiler Feed Water Pressure Regulator (Courtesy of Watts Regulator Company)

Maintenance Sediment from the domestic water supply will enter the water feed piping and obstruct flow over time (Figure 9-47). This problem should be dealt with in accordance with manufacturer's instructions. In addition, maintenance intervals should be adjusted by the actual conditions observed during past service events.

FIGURE 9-47 Scale Fouling of Water Feed Valve (Courtesy of Watts Regulator Company)

Also, the quality of water may differ at different locations and will affect the frequency of needed maintenance intervals. Homes on well systems, for example, may have a poorer water quality than municipally provided domestic water and thus require more frequent service. Also be alert to recent water main flushing or repairs since either of these occurrences may dislodge sediment in the system, interfering with the operation of the water supply's auxiliary components.

Routine preventative maintenance or manufacturer-recommended service should include removal, inspection, and cleaning of any internal strainer screening that may be blocked by debris (Figure 9-48). If any parts seem worn or the strainer screen seems excessively fouled to the point that cleaning may not be completely effective, the old strainer assembly should be replaced by a new assembly (if available).

FIGURE 9-48 Internal Strainer Screen (Courtesy of Watts Regulator Company)

Pressure-regulating valves also need to be checked. These valves rely on an internal synthetic rubber diaphragm to sense system pressures. Over time and as a result of water impurities, these diaphragms may become damaged and leak (Figure 9-49). Signs of leakage, including water leakage from the regulator valve body, indicate that the diaphragm is leaking and should prompt a replacement of the diaphragm or valve (Figure 9-50).

FIGURE 9-49 Water Leakage from Valve Diaphragm (Courtesy of Watts Regulator Company)

FIGURE 9-50 Inspecting Valve Diaphragm (Courtesy of Watts Regulator Company)

Consequences of Failure An interruption in water flow to a boiler can result in air being entrained in the system water, causing baseboard and hydronic system knocking and water hammer as well as overheating and thermal shock damage to the boiler castings and heat exchangers. Care must be taken to make sure that the stop valves are turned back on upstream of the water feeder and the backflow valves after service has been completed and the system has been returned to operation.

Forced Hot Water Boiler—Backflow Preventer

Although most pressure-reducing valves have an integral check valve, nondrinkable boiler system water is prevented from flowing back into the domestic water source with the use of a backflow-preventing check valve designed specifically for this purpose. A better way to state this is to say that a backflow preventer is a more reliable check valve as opposed to a simple check valve that, although it performs the same rudimentary task, is not as reliable.

Description Backflow preventers have multiple check valves within their construction as well as strainer screens to prevent foreign contaminants from affecting their operation.

Maintenance Backflow preventers should be inspected, tested, and cleaned in accordance with the manufacturer's recommendations. These inspections may be as frequent as every 6 months.

Maintenance includes removing the device and cleaning any accumulated sediment (Figure 9-51) that has found its way to the valve since the last service event. Manufacturers also provide testing procedures for backflow preventers to assure that they are operating properly after installation and subsequent maintenance actions.

FIGURE 9-51 Fouling of Backflow Preventer Valve (Courtesy of Watts Regulator Company)

Consequences of Failure An obstructed backflow preventer, like an obstructed water feeding valve or even a closed valve in the boiler supply line, can cause the boiler to run low on water, overheat, and suffer thermal shock damage.

If sediment finds its way into the backflow preventer and interferes with the internal check valve's ability to reseat, this can cause a continuous uncontrolled flow of water from the boiler water supply line, causing water damage to the boiler as well as the mechanical space where it is installed.

Steam Boiler Water Feeder

A steam boiler water feeder valve automatically adds water a steam boiler. Without this valve, water would have to be added manually, which, if a property is vacant, could allow the system to shut down due to the lack of water in the boiler.

A water feeder valve on a steam boiler may be part of a low-water cutoff switch or may be a separate add-on control (Figure 9-52).

FIGURE 9-52 Steam Boiler Water Feeder Valve
(Courtesy of McDonnell & Miller–Xylem Inc.)

What is important to remember is that the water feeder relies on the signal the boiler's low-water cutoff switch is sending to it. If, for example, the low-water cutoff switch is not maintained and fails to send a low-water signal, the boiler will overheat, and likewise the water feeder valve will never get the input needed that will cause it to add water to the boiler.

Just as a low-water cutoff switch can fail to detect a low-water condition, it is also possible that it cannot detect a high-water condition. In this case the water feeder valve could continually send water to the boiler, overfilling it as well as potentially filling the entre steam system piping with water!

Maintenance The boiler's low-water cutoff switch, like all other controls, should be inspected during annual (or more frequent) preventative maintenance intervals. Care must be taken to identify any wear or corrosion-damaged components, and any internal strainers should be inspected and cleaned as needed. Also it is important to make note of the control's age since manufacturers may specify a service life of the device.

Air Vents and Steam Traps (Steam and Forced Hot Water Systems)

Air Vents

One of the problems with steam and forced hot water systems is that entrained air may impede the flow of steam and interfere with the return of condensate to the boiler. Air vents on steam radiators (Figure 9-53) allow the removal of air from steam system piping as the piping and radiators become filled with steam. Air vents are present on both steam and hot water systems and these should be inspected to prevent unwanted water leakage from the system.

FIGURE 9-53 Steam Radiator Air Vent

Maintenance With the system in operation, look for a constantly discharging air vent. A vent that is constantly emitting or dripping steam (Figure 9-54) is likely letting steam and water escape from the system in amounts that may affect the water chemistry of the boiler. This is an indication that the vent has failed and needs replacement.

On steam systems, use a noncontact thermometer or infrared imaging to look for nonuniform heat distribution through the radiator. If a radiator heats in an uneven manner, this is an indication that there is a problem with the air vent. It likely means that the upper region of the radiator is filled with air and that the steam vent is not operating or is obstructed, preventing the removal of air from the system. The solution again is to replace the vent. Indicators of leaking forced hot water air vents include corrosion and hard mineral deposits.

FIGURE 9-54 Leaking Steam Radiator and Forced Hot Water Air Vents

Air vents are relatively inexpensive, and considering the performance benefits obtained from retaining steam that could be wasted, the maintenance and needed replacement of these components of the heating system is a worthwhile investment.

Consequences of Failure Due to steam losses from vent failures and other causes, excessive makeup water must be added to a steam boiler. Over time, poor boiler water chemistry issues arise. One consequence of excessive makeup water in steam boilers is corrosion hole damage above the waterline (Figure 9-55).

FIGURE 9-55 Corrosion Hole in Boiler Section

Steam Traps (Steam Systems)

There are different types of steam traps. One type is the float trap, which is usually seen in larger or commercial system applications. The type most often seen in residential steam system applications is the bellows or balanced pressure trap (Figure 9-56). The bellows type is the focus here.

FIGURE 9-56 Steam Trap on Cast Iron Radiator

Description The bellows type of control relies on an internal bellows cartridge that is filled with a reactive mixture of water and mineral spirits that either vaporize or condense at the water-steam phase point. In reaction to this, the bellows either contracts to open the steam trap valve to allow condensate to leave the system or expands to close the steam trap valve to keep the steam in the trap until there is enough condensation to reverse the process (Figure 9-57).

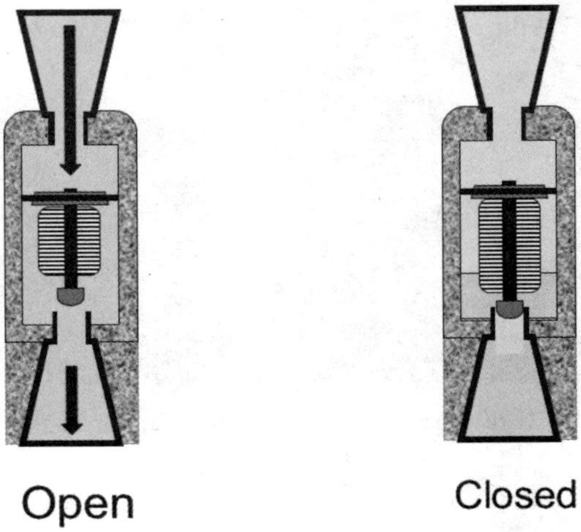

FIGURE 9-57 Bellows Steam Trap Operation

Maintenance Steam traps fail over time due to mechanical wear and tear as well as rust and piping contaminants interfering with their operation. These problems will cause the trap valve to become obstructed or not seat entirely, causing the radiator to leak steam. Only condensate should come out of the radiator trap and not steam.

A good way to identify if a steam trap has failed is to use an ultrasonic leak detector. This detector will pick up the sound of steam leaking through a steam trap that is supposed to retain the steam until it turns back into water (condensate).

Oftentimes failed steam traps can be repaired by replacing the bellows assembly (Figure 9-58).

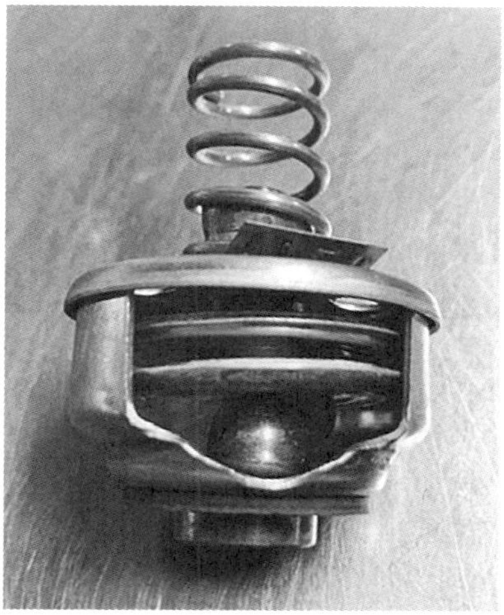

FIGURE 9-58 Replacement Steam Trap Bellows Assembly

Consequences of Failure In the event of a steam trap failure, steam will continually pass through the radiator as opposed to being retained until its latent heat of steam vaporization is expended through it to the room. This will cause accelerated fuel consumptions and inefficiencies. Also the introduction of steam back into the system piping intended for condensate will move this condensate at velocities not intended, resulting in steam hammer, or knocking sounds, from the piping system.

Control Parameter—Combustion

In the combustion appliance, it is a vital necessity to control the sequencing and operation of the burner system, whether or not the fuel is oil, natural gas, or propane. In the event of a flameout, the possibility of a fuel accumulation of any type must be avoided. An unburnt oil flow into a boiler or furnace will flood the appliance with oil, rendering the appliance in need of extensive repair or replacement. In a gas-fired appliance, a gas leak is poisonous to the occupants of the building. And on another level, the ignition of any accumulated fuel results in uncontrolled combustion and explosions.

Preventative maintenance should include checking out and testing the appliance's combustion control to verify that it is capable of shutting down the appliance in the event of a flameout. The maintenance check also must verify that the control will not allow fuel to continue to flow after a flameout and that it will operate safely in accordance with the manufacturer's intentions.

There are different controls and different control operational principles that make up each combustion control design. The control should be able to react to the foreseeable possibilities of unintended operation and to shut down the appliance in a safe and reliable manner. That ability can be re-created by the technician performing maintenance to verify that the control is capable of what is expected.

The Oil Burner Primary Control

The CAD (cadmium sulfide) cell control is the oil burner primary control and is the basis of all modern combustion control relay devices.

Description The CAD-cell primary control uses a light-sensitive CAD photo eye positioned in the oil burner blast tube to verify the presence of a flame upon the initiation of combustion. When the oil burner is energized and combustion begins, the CAD-cell eye (see the arrow in Figure 9-59) senses the burner flame in the combustion chamber, allowing the burner to be continually energized in response to a call for heat. In the event that there is no or a poor quality of flame, the primary control in a residential oil burner will shut down the burner's power circuit after the control's particular proving time is exceeded.

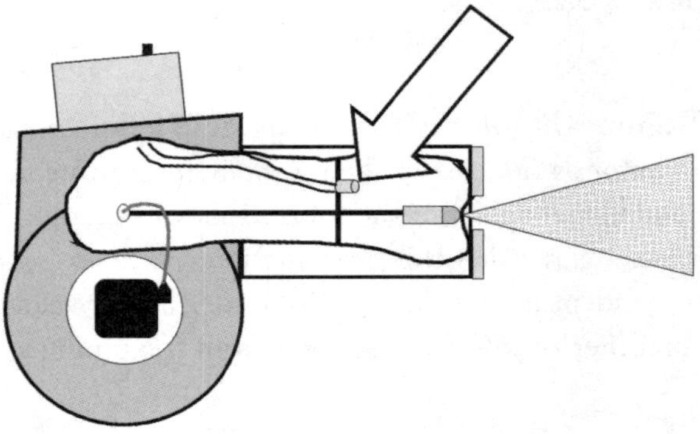

FIGURE 9-59 CAD-Cell Eye Positioned in Oil Burner

Maintenance The primary control is usually mounted on the burner body (Figure 9-60), and servicing the burner requires manipulation of the ignition transformer and other actions that may impart tension to the wiring connections. As well, the various wiring connections under the primary control electrical enclosure can be compromised. And another potential problem is that there are too many wires to keep track of and the wrong wiring nut is used, creating a poor connection (Figure 9-61).

Following a burner service, a simple and easy maintenance task to perform is to verify if all the wiring connections are tight.

FIGURE 9-60 CAD-Cell Primary Control (Courtesy of Honeywell–Resideo)

FIGURE 9-61 Examining Wiring Connections to Primary Control

During maintenance, the ignition transformer should also be tested for failure. You can test the control for its ability to shut down or "lock out" the burner by starting the burner and then removing the CAD-cell eye's electrical connections from the control to simulate a "no-flame" condition. The burner should shut down.

The manufacturer's specifications for the CAD-cell eye's resistance values in both a "light" and "no-light" condition should be verified as well.

The Gas Valve and Flame Verification System

No matter whether the gas burner relies on a flame rod or a pilot light–thermocouple arrangement, the gas valve can be tested to see if it will actually respond to a flameout condition and shut the appliance down.

Wall-Mounted Boilers and Appliances with Electronic Controls

A recent trend in heating system manufacturing is appliances with electronic controls that also have diagnostic capabilities. Initially this trend began with the aftermarket addition of furnace control boards and ignition modules. It has now progressed beyond these to manufacturer-specific components that cannot be substituted with aftermarket components or electronic controls of a generic nature.

No maintenance of the controls is needed from an operational point of view. However, visual observations of the control boards can potentially identify an impending problem that may later result in appliance damage and shutdown.

It is important to be aware of issues such as internal condensate or water leakage, which can directly damage electronic controls as well as create a high-humidity environment that can damage them as well. And as is the case with any appliance, especially one installed in a location accessible to the exterior, a warm appliance is a desirable nesting location for rodents, which leads to waste products being deposited in the appliance that are corrosive and damaging to these controls.

Water Contact

Be alert to situations where a boiler control has come into contact with water due to condensation in high-efficiency gas appliance condensate leakage or due to boiler water leaks from air vent gaskets or other component leakage. Water entering the controls can cause a stray unintended electrical current that may defeat the intended functions of the integrated circuit components. Water entry into electrical connections causes interface oxides to form, reducing the integrity of the electrical connection, which in turn creates a heat-generating connection capable of triggering electrical fires in the board and control (Figure 9-62).

FIGURE 9-62 Damage to Circuit Board Due to a High-Resistance, Heat-Generating Connection

TABLE OF APPLICABLE PREVENTATIVE MAINTENANCE CONTROL ACTIONS

Control Type	Steam Boilers	Forced Hot Water Boilers	Forced Hot Air Furnace
Float type of low-water cutoff switch	X		
Probe type of low-water cutoff switch	X	X	
Safety relief valve	X	X	
Pressure control	X		
Expansion tank		X	
Aquastat		X	
High limit—fan			X
Steam—air vent	X		
Steam trap	X		
Primary control (oil)	X	X	X
Gas valve	X	X	X

CHAPTER 10

COMBUSTION CHAMBER REFRACTORY

Boilers and furnaces are constructed with high-temperature insulation (refractory) that insulates the area that the burner flame occupies. This area is known as the firebox or combustion chamber, and the refractory is necessary due to the temperatures developed, which will exceed the melting point of steel (Figure 10-1). Without this refractory in its proper place, the metal of the heat exchanger or boiler components can overheat, causing the products of combustion to escape from the appliance.

FIGURE 10-1 Combustion Chamber Refractory and Depiction of Oil Burner Flame

Additionally, soot production and interference with combustion product flow often result due to a fallen or improperly positioned refractory, which can cause blockage of the appliance flue passageways and the introduction of products of combustion (soot) into the indoor environment.

There is a strong need for the service technician to assess the condition of this refractory on a preventative maintenance schedule before the refractory finally fails. This chapter discusses the causes of refractory failure and the visual indicators of impending refractory failure.

Preventative and Reactive Maintenance

A heating company's service plans include many maintenance actions. These maintenance actions are designed to make sure that a combustion appliance operates safely and efficiently and for as long as possible. As first noted in Chapter 1, maintenance actions can be reactive, preventative or predictive:

- **Predictive maintenance** is assessing what maintenance will be required in the future. An example of this is ordering parts for and scheduling the repair of an appliance based upon known wear rates or life expectancy of parts.
- **Preventative maintenance** is maintenance that is performed to prevent a more major catastrophic failure. An example of this is the annual replacement of oil filters and nozzles on an oil burner system.
- **Reactive maintenance** is maintenance that is performed after a failure occurs. An example of this is clearing a boiler that is plugged with soot or fixing a leaking gasket.

Preventative maintenance actions are not limited to the work that required a part replacement or maintenance action only. A crucial element of preventative maintenance is the visual assessment of the condition of the appliance. The *visual inspection* is vital to identify the need for replacement of certain components *before* failure occurs.

Additionally, experience with past failures of similar components must be incorporated in preventative maintenance actions with the best judgment that the service technician can apply to prevent an unexpected appliance failure.

Combustion Chamber Refractory

The temperature of combustion can exceed the melting point of steel and cast iron, the materials of which boilers and furnace 's fireboxes are constructed with.

Combustion chambers are capable of also acting as a "heat sink" or "reservoir of heat" to provide additional energy to the medium being heated (air in a furnace or water and steam in a boiler), allowing longer periods between firings of the burner during heating operation.

The combustion chamber refractory has been made of a number of materials over the years, and a service technician may encounter a number of these different designs and materials. As such, unless your service work is limited to only new appliances, it is a good idea to familiarize yourself with what you may encounter.

Past Materials and Designs

Historically, combustion chambers were fabricated during either the appliance installation or the burner retrofit.

Firebrick

Firebrick (Figure 10- 2) was one of the first refractory materials used. Firebricks have an upper melting point between 1600° and 4000°F and are made from a number of materials including fire clay (kaolin), silica, alumina, and magnesite. Additionally, firebrick was sometimes made with asbestos fibers embedded in the insulating medium.

FIGURE 10-2 Combustion Chamber Firebrick

Firebricks are still available in a number of shapes and sizes and can still be used for construction or repair of residential and older commercial installations. High-temperature mortar is used to bond and seal firebrick components.

When clay or kaolin firebricks undergo failure, they go through a deterioration process where the brick surface crumbles and develops small areas of deterioration known as spalling. Eventually the brick and adjoining high-temperature mortar will begin to crumble and become dislodged, causing the brickwork to fail (Figure 10-3).

FIGURE 10-3 Failing Boiler Firebrick

In the construction of the firebrick combustion chamber, anchor points in the boiler or furnace casing may be installed to secure the mortar to the metal chamber. Also, the insulating firebrick may be installed over a second level of insulation material, such as sand, vermiculite, or other refractory brick layers, between it and the appliance walls.

Variations of field installations also exist, consisting of a mixed or a combination of brick and mortar combustion chamber constructions.

In the example in Figure 10-4, the chamber consists of an inner core of firebrick that was parged over with a skim coat of high-temperature insulating refractory cement either during its initial installation or during a repair at some time in the past. Note the vermiculite fill material between the boiler's heat exchanger and the combustion chamber refractory.

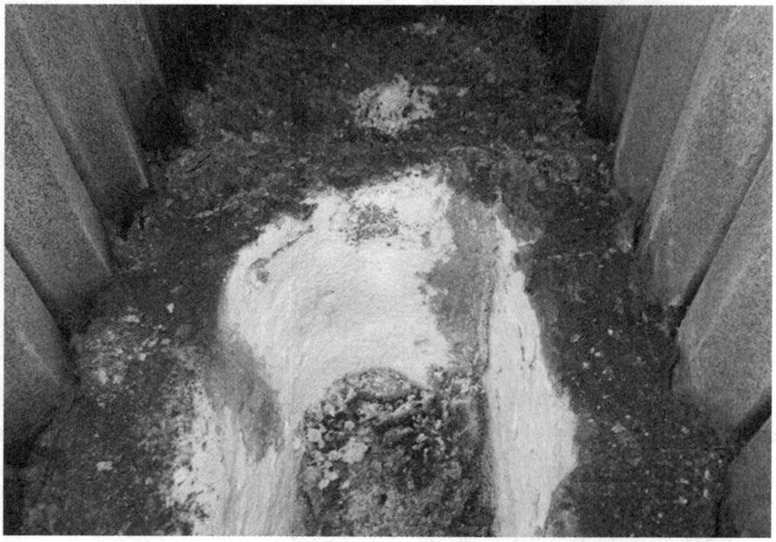

FIGURE 10-4 Combustion Chamber Made of Both Brick and Mortar

Molded Refractory

Harder clay-like fireboxes were common in boilers and furnaces in the late 1950s and early 1960s and went by the name of Lib-O-Rator Molded Refractory Combustion Chambers (Figure 10-5). These had good durability, but they were brittle and prone to cracks in the event of a delayed ignition explosion.

FIGURE 10-5 Molded Refractory Combustion Chamber Refractory

Steel Combustion Chambers

Some oil-fired furnaces are outfitted with combustion chambers made of steel (Figure 10-6). Care must be taken with these chambers to adhere to the manufacturer's installation and burner setup instructions to avoid any conditions contributing to reduced airflow over the chamber or to flame impingement. Reduced airflow and flame impingement can both result in localized overheating of the metal firebox chamber inside the heat exchanger, leading to failure.

FIGURE 10-6 Steel Combustion Chamber

The initial stages of metal chamber failure are evident by the warping and melting of the steel chamber. Although a patch of ceramic fiber refractory can be added for a short-term remedy, since the integrity of the steel chamber's structure can be affected by continual deterioration, replacement of the chamber should be considered provided that replacement components are available.

Ceramic Fiber Combustion Chamber Refractory

Today the use of ceramic fiber combustion chamber refractory is widespread. Ceramic material comes in rigid, blanket, and loose fill; and it also comes in a liquid-impregnated form, allowing it to be molded into different shapes and custom-fit for odd-sized or quick repair of failed refractory fireboxes. These refractories can withstand firebox combustion chamber temperatures and are available in shapes that allow any appliance's combustion chamber to be adequately protected.

Chamber design and size must be coordinated with the flame pattern of the oil burner equipment, mainly in the selection of nozzle size and pump pressure. Sizes and shapes of ceramic fiber refractory "kits" are constructed to fit any number of furnace and boiler chamber designs. See Figure 10-7.

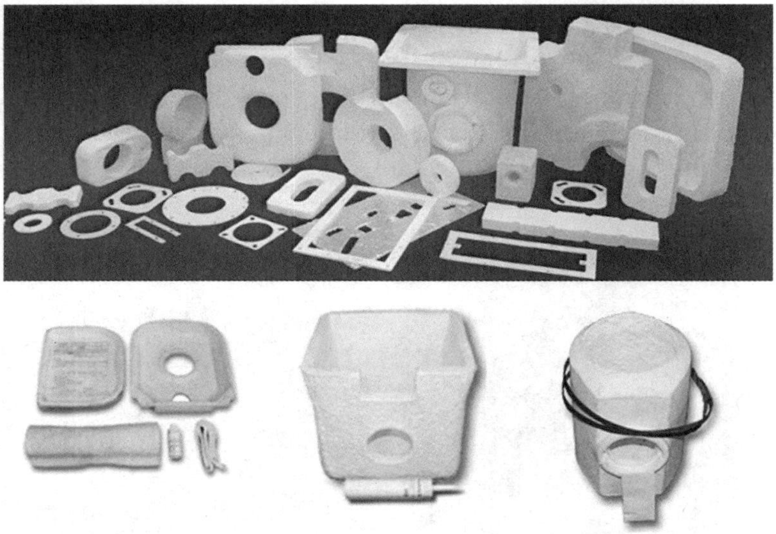

FIGURE 10-7 Various Combustion Chamber Refractory Geometries (Courtesy of Lynn Manufacturing Inc.)

Types of Refractory Designs for Types of Appliances

The shape of the combustion chamber refractory corresponds to the shape of the boiler's or furnace's firebox, which is the portion of the appliance that receives the refractory combustion chamber material.

For oil- or power burner–fired appliances, the shape and dimensions of the combustion chamber are interrelated with the setup of the oil burner and oil nozzle selection. As such, in selecting the proper refractory pieces or repairs (temporary or not) when making repairs to the combustion chambers of older equipment, the technician must take into account the intended design size and shapes of the original chamber.

As stated in Chapter 6 on oil burners, the size of the nozzle, the nozzle flow rate, the spray angle, and the shape of the cone, along with the fuel pump pressure, determine the "oil cone" and resulting flame geometry (Figure 10-8). Nozzle sizes and matching specifications for installation into a particular appliance are typically listed by boiler or furnace manufacturer. However, for older appliances that have been retrofitted with oil burners, reverse engineering to determine the nozzle size must be performed.

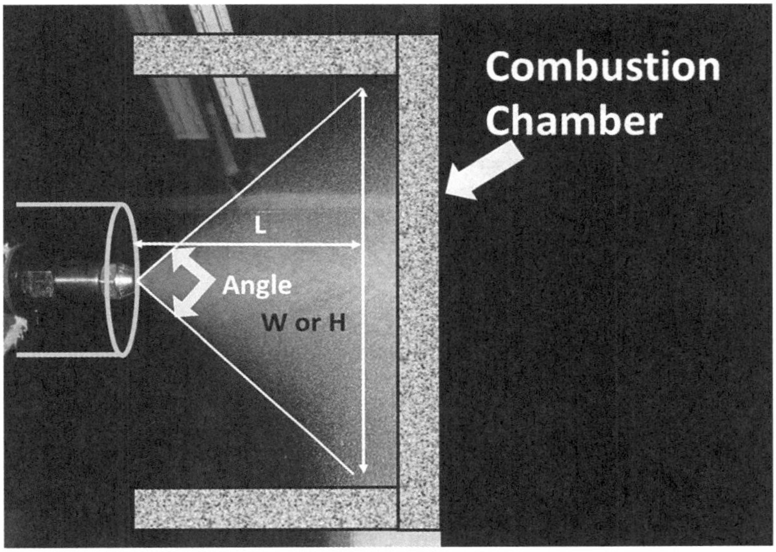

FIGURE 10-8 Positioning of the Refractory Within the Dimensions of the Combustion Chamber Is Crucial for Proper Operation

By identifying the dimensions and geometry of the appliance's combustion chamber, the appropriate nozzle and pump pressure that will produce an appropriately sized flame can be identified. It is important that the burner flame does not exceed the boundary of the combustion chamber. If it does, the burner flame will strike the chamber walls (known as flame impingement), resulting in soot production and accelerated combustion chamber refractory failure.

Oil-Fired Forced Hot Air Furnaces—Cylindrical

Oil-fired forced hot air furnaces are commonly outfitted with a cylindrical combustion chamber positioned either vertically or horizontally (Figure 10-9). The chamber may be positioned on the heat exchanger floor or supported with a metal frame to position it within the heat exchanger and oil burner input (Figure 10-10).

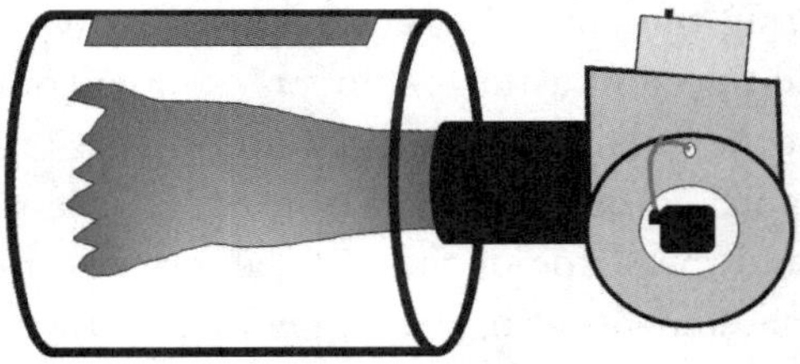

FIGURE 10-9 Cylindrical Combustion Chamber

FIGURE 10-10 Cylindrical Combustion Chamber Removed from Furnace

With oil-fired furnaces especially, extra care must be exercised to make sure that the combustion chamber is intact. If it is not, this will result in heat exchanger damage if the refractory fails.

Depending on the furnace design, access to the combustion chamber for inspection may require removal of the oil burner to allow visual access through the oil burner air tube opening or mounting "pouch plate" (Figure 10-11).

The procedure for replacing the combustion chamber requires disassembly of the furnace including the burner, jacket sections, and combustion chamber access plates. Due to the heat generated in the chamber, it is likely that bolts and other fasteners have seized, requiring extra effort to remove or replace.

In addition to refractory replacement, gaskets and seals may be torn during the disassembly, so it is a good idea to research the parts replacement list of the appliance you are working on so that all replacement refractory pieces as well as gaskets and seals are on hand.

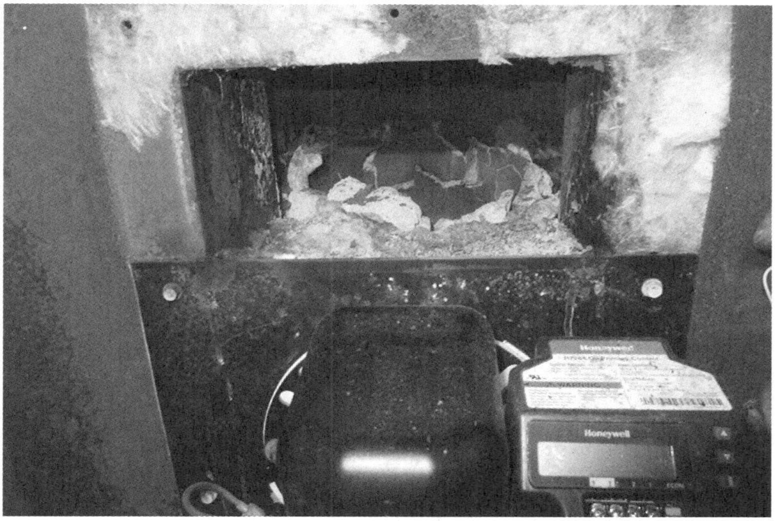

FIGURE 10-11 Removed Oil Burner and "Pouch Plate" Allowing Access to Combustion Chamber Refractory for Visual Inspection

Oil-Fired Cast-Iron Wet Base Boiler

For an oil-fired cast-iron wet base boiler, the main components of the combustion chamber refractory consist of the rear or target wall, floor blanket, and front or door refractory pieces (Figure 10-12). Inspection of the refractory can be done by removing the combustion chamber—oil burner mounting plate—as well as through the chamber inspection door as well as through the oil burner air tube opening after the burner is removed (Figure 10-13).

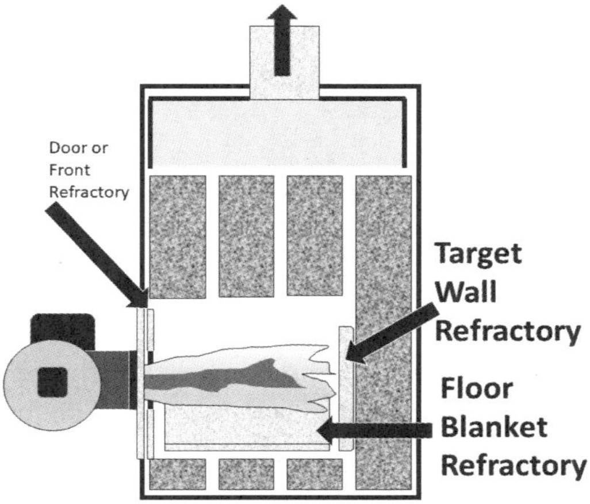

FIGURE 10-12 Positioning of Combustion Chamber Refractory in Cast Iron Wet Base Boiler

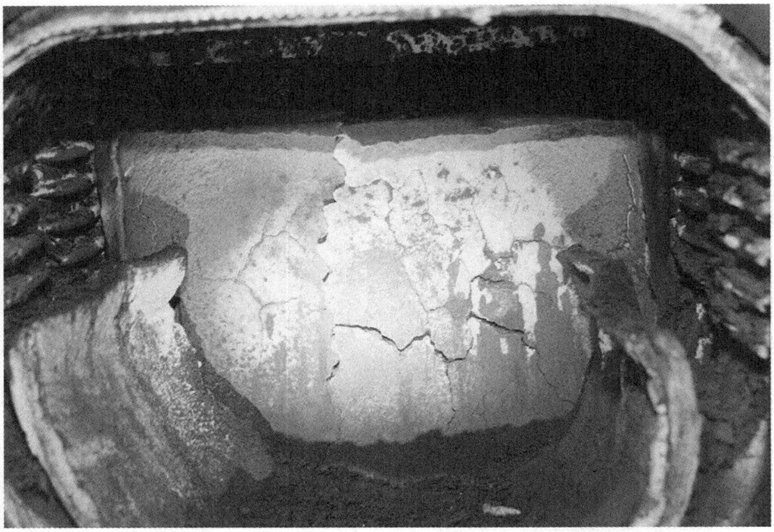

FIGURE 10-13 Burner Mounting Door Removed Revealing Combustion Chamber Refractory

MAINTENANCE TIP!

Chamber Replacement . . . Keep It Going!

Replacing the combustion chamber refractory of a boiler before the refractory fails completely will prevent damage to the appliance as well as prevent damaging soot discharges into the home. Replacing the combustion chamber refractory is part of the expected maintenance procedures, and if done properly and between times of ultimate combustion chamber refractory failure, it can prolong the appliance's continued use likely beyond its expected lifetime.

Replacement of the Combustion Chamber Refractory of a Cast-Iron Wet Base Boiler

The first step in the replacement of the combustion chamber is to remove the oil burner and combustion chamber door while at the same time vacuuming and clearing the boiler's firesides and combustion chamber of accumulated soot, pieces of the refractory, and other debris (Figure 10-14). This entails cleaning all components of the refractory including the front door piece (Figure 10-15) as well as floor blanket, target wall, and gasketing material (Figure 10-16).

FIGURE 10-14 Burner Mounting Door Removed Exposing Combustion Chamber Refractory (Courtesy of Lynn Manufacturing Inc.)

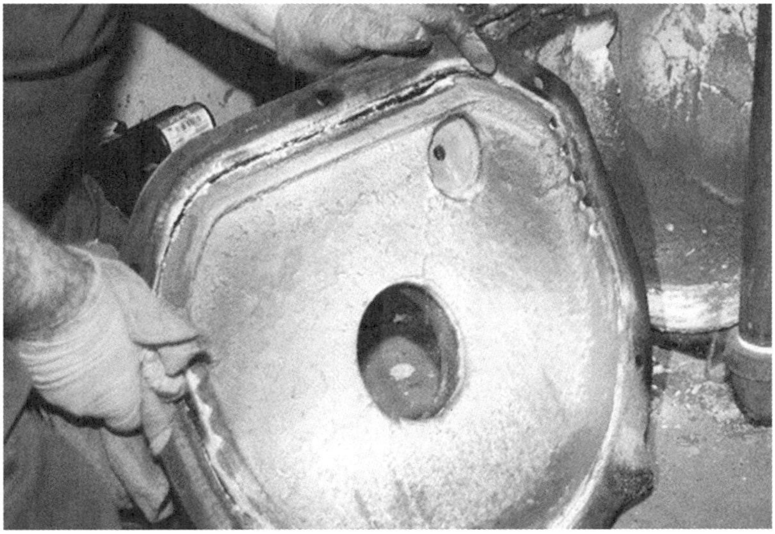

FIGURE 10-15 Refractory Front Door Piece
(Courtesy of W.H. Riley Oil Company and Lynn Manufacturing Inc.)

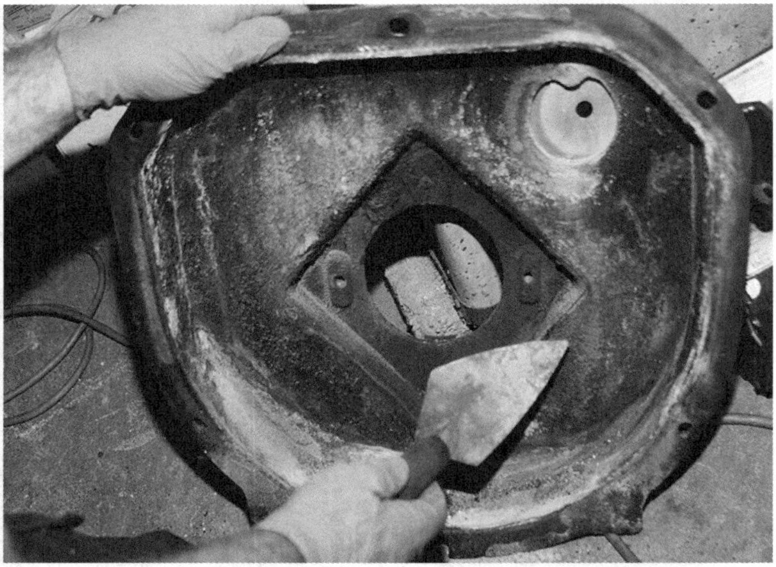

FIGURE 10-16 Refractory Removed from Front Door (Courtesy of W.H. Riley Oil Company)

For refractory components that need to be secured in place by an adhesive, water glass, or sodium silicate is used (Figure 10-17). The water glass is applied to surfaces where gravity or resistance between the boiler components and refractory pieces alone will not secure the refractory pieces in place. The water glass is applied to clean surfaces, and then the refractory components are pressed into place (Figure 10-18 and 10-19).

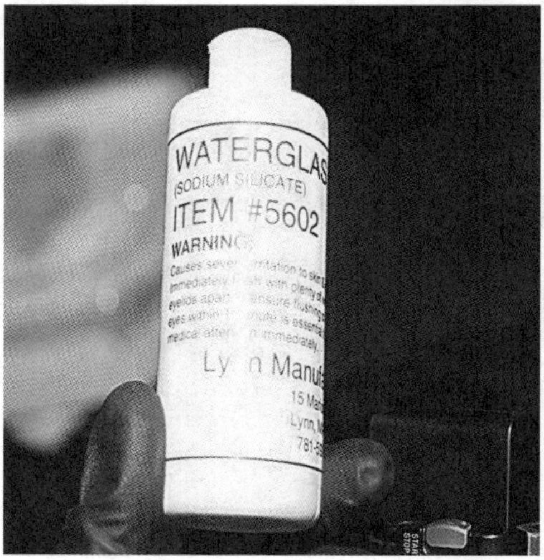

FIGURE 10-17 Water Glass (Courtesy of Lynn Manufacturing Inc.)

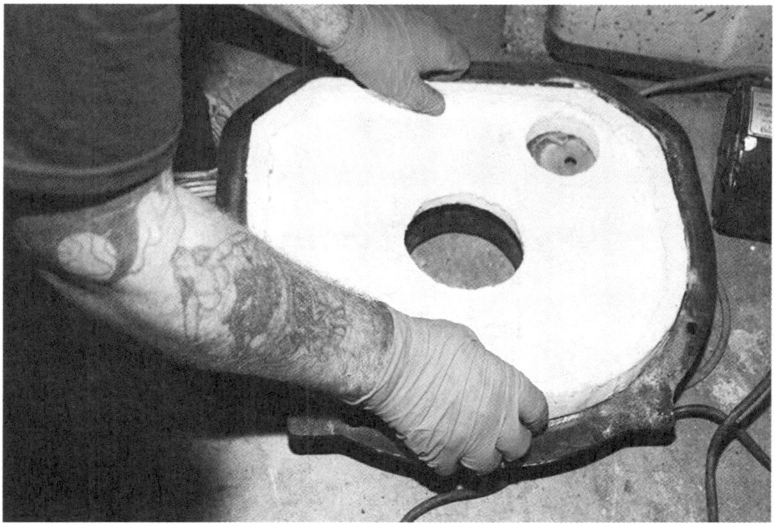

FIGURE 10-18 Reinstalling Front Door Refractory Piece After Water Glass Adhesive Applied (Courtesy of W.H. Riley Oil Company and Lynn Manufacturing Inc.)

FIGURE 10-19 New Combustion Chamber Refractory in Place (Courtesy of W.H. Riley Oil Company and Lynn Manufacturing Inc.)

Oil-Fired Steel or Cast-Iron Dry Base Boilers

In the dry base boiler, the firebox is dependent solely on the integrity of the refractory. It is this same refractory that protects the firebox from the heat of the products of combustion and prevents the escape of products of combustion to the outside of the appliance.

The combustion chamber of the dry base boiler is a rectangular refractory structure that sits inside the boiler's firebox and beneath the heat exchanger (Figure 10-20).

To replace the refractory, the boiler must be disassembled. This requires that the jacket be removed, the steel heat exchanger be either supported or removed, and the firebox be accessed (Figure 10-21).

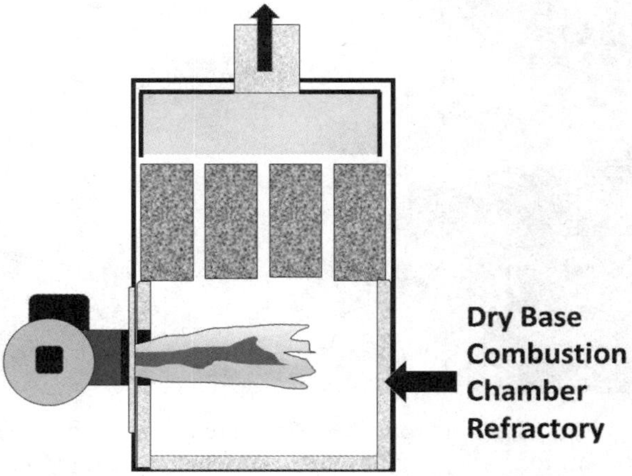

FIGURE 10-20 Dry Base Boiler's Combustion Chamber Refractory

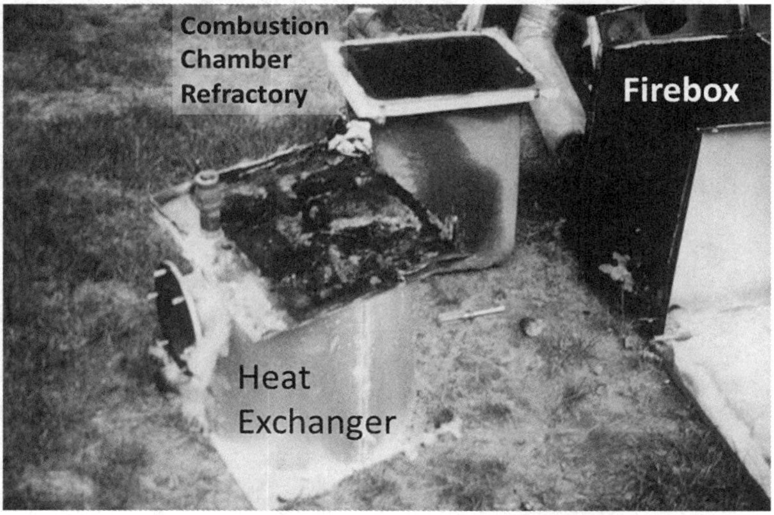

FIGURE 10-21 Dry Base Boiler's Combustion Chamber Refractory Removed from Firebox

Gas-Fired Cast-Iron Boiler

On standard gas-fired boilers, ceramic fiber refractory panels line the perimeter of the combustion chamber (Figure 10-22). The refractory panels are positioned beneath the heat exchanger and above the gas burners and are easily accessed for inspection by removing the gas burners and boiler rollout shields (Figure 10-23).

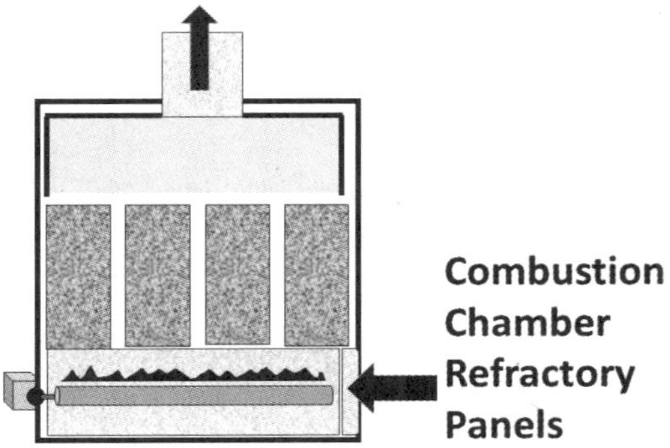

FIGURE 10-22 Gas-Fired Boiler Showing Position of Combustion Chamber Refractory Panels

FIGURE 10-23 Refractory Panels Inside of Gas-Fired Boiler's Combustion Chamber

The refractory panels are held in place with insulation fasteners (Figure 10-24), or the insulation pieces can be inserted into tabs that hold them in place aligned with the gas burners, as seen in Figure 10-25.

FIGURE 10-24 Insulation Fasteners

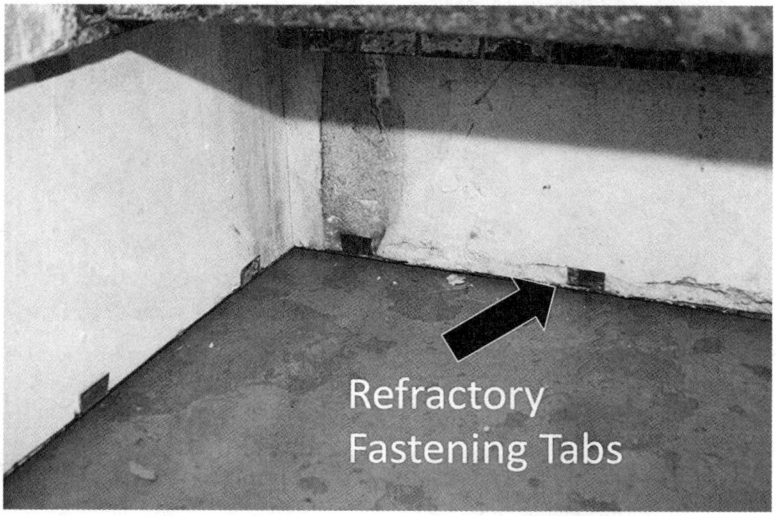

FIGURE 10-25 Insulation Fastening Tabs

Causes of Failure

The chamber refractory is damaged by the effects of erosion, continual expansion and contraction, and fatigue. Additionally, water contact will affect the integrity of the refractory.

The environment in which the refractory operates is one that fluctuates from ambient to extreme temperatures, and this also leads to failure. A burner stopping and starting will deliver a high-velocity flow of hot combustion gases over the surface of the refractory thousands of times in a few heating seasons. This results in expansion and contraction of the refractory material as well as erosion as hot combustion gasses pass over the refractory surfaces.

The chamber refractory must maintain its thickness in order to insulate properly, and it must maintain its rigidity in order to retain its structure and stay in place. As time as well as the effects of expansion, contraction, and erosion continue, this thickness is reduced resulting in a reduced insulation capability as well as diminished insulation structure integrity, making it prone to collapse.

How long should a combustion chamber refractory last? Various factors affect the life of a combustion chamber refractory. These include initial design, burner setting and adjustment (flame impingement quickly erodes the refractory), and nozzle size. Some chambers have seen over 20 years of service, and others fail in as short a time as 7 years.

FORENSIC INSIGHT

Signs of Impending Failure

Visual Inspections—Detecting Impending Failure

The life span of a chamber refractory structure can vary from appliance to appliance. To identify whether a refractory needs to be replaced before total failure occurs, look for the following problems during service intervals to predict the ability of the refractory to survive the upcoming heating season.

Surface Cracks

Surface cracks are the first indication that the ceramic fiber chamber is deteriorating. These cracks are often concentric in pattern and more numerous on the target wall area (Figure 10-26).

FIGURE 10-26 Surface Cracks in Combustion Chamber Refractory

For gas-fired boilers, the refractory panels will also show signs of wear and tear and shrinkage, as seen in Figure 10-27. This damage may be accompanied by signs of product-of-combustion leakage on the boiler's exterior as hot combustion gases leak out of the seal area of the refractory-boiler heat exchanger and escape into the space between the boiler and sheet metal jacket.

FIGURE 10-27 Refractory Panels in Gas-Fired Boiler Showing Leakage Between Chamber and Heat Exchanger

Curling/Shrinking/Sagging/Crumbling

As refractory fibers break due to continual expansion and contraction, the structural integrity of the refractory shape will begin to fail. Warping and shrinking or sagging of the chamber will result, which is a precursor to the chamber falling apart entirely (Figure 10-28).

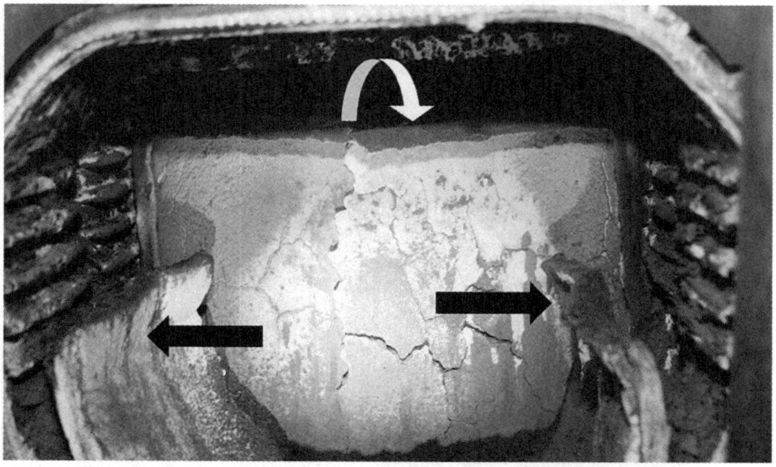

FIGURE 10-28 Impending Failure of Oil-Fired Boiler's Combustion Chamber Refractory

In addition to sagging, the edges of the refractory pieces may start to crumble (Figure 10-29), with pieces falling into the lower areas of the combustion chamber.

FIGURE 10-29 Crumbling Edges of a Round Combustion Chamber

Thinning and Spalls

Thinning of the refractory is an indication that the remainder of the refractory's useful life is limited. In the event that the thickness of the refractory is reduced significantly, the refractory should be replaced (Figures 10-30 and 10-31).

FIGURE 10-30 Comparing the Combustion Chamber's Thickness in Wear Areas of the Chamber to that of its Original Thickness

FIGURE 10-31 Over Time the Combustion Chamber's Thickness Has Become Thin

In inspecting the refractory for thinning, pay particular attention to the rear wall (target wall) area of the chamber. In Figure 10-32 the target wall's refractory was "paper thin." It's no wonder the furnace had a hole in the heat exchanger.

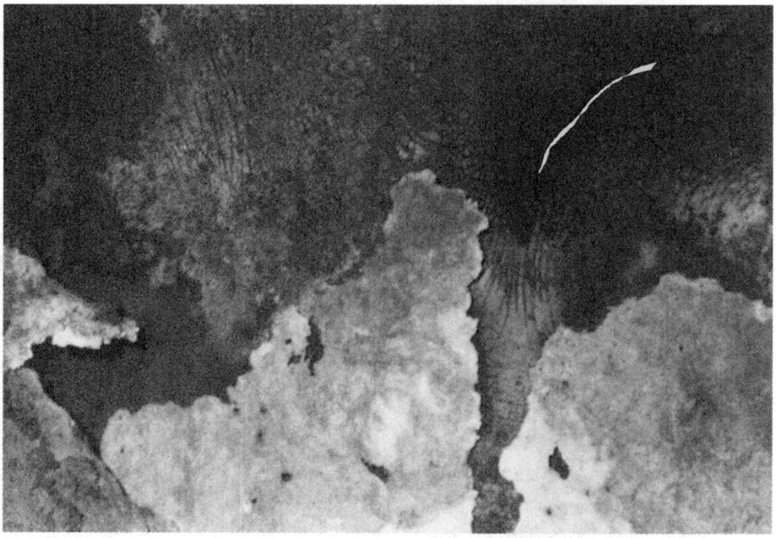

FIGURE 10-32 Reduction of Combustion Chamber Refractory to the Point Where It Has Become Paper Thin

Also, if a "crater-like" depression of any depth (a spall) is seen in the refractory, this is an indication that the flame is impinging on the liner and that the refractory should be replaced.

Dry Base Boiler Refractory Failure Indicators

Dry base boilers depend on the condition of the combustion chamber refractory since its integrity is the only component containing combustion gases (Figure 10-33). An early indication of dry base boiler chamber failure is the leakage of combustion gases at the gap/seal between the combustion chamber and the boiler's heat exchanger (Figure 10-34).

FIGURE 10-33 Interface Seal Between Firebox and Heat Exchanger in Dry Base Boiler

FIGURE 10-34 Leak at Refractory Seal Area in Dry Base Boiler's Combustion Chamber

Cleaning Tool Damage

A service technician's duties should include a visual examination of the refractory to determine that it has not been damaged by cleaning tools and is capable of continued service. Damage to the refractory can result from cleaning instruments striking the chamber (Figure 10-35).

FIGURE 10-35 Cleaning Tool Damage to Combustion Chamber Refractory

When a cleaning brush is inserted into a boiler's or furnace's heat exchanger flues, care must be taken to ensure that it does not extend so far into the appliance that it enters the combustion chamber area and damages the refractory (Figure 10-36). As well, care must be taken to make sure that the liner is not damaged during service calls.

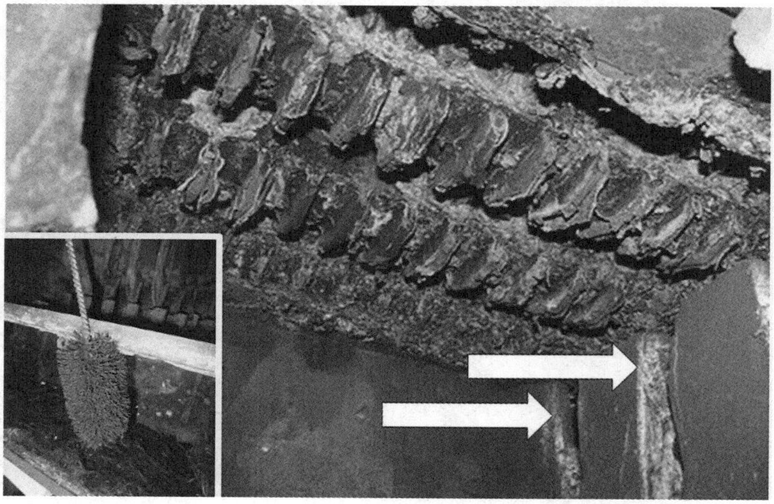

FIGURE 10-36 Brush Passage into Combustion Chamber Should Not Extend So Far as to Strike Refractory Damaging It

Water Contact

The various refractories are susceptible to water contact to varying degrees.

Refractory material that is saturated with water—caused perhaps by a flooded basement (Figure 10-37), or a flooded mechanical room, or a leaking pipe—may cause the refractory to become dislodged or fall apart under the weight of the water inside it.

FIGURE 10-37 Submerged Boiler Due to a Basement Flood

It is also possible that the refractory is becoming wet due to an unknown leak from the heat exchanger push nipple or gasket (Figure 10-38).

FIGURE 10-38 Leaking Boiler Section Is Allowing Water to Contact Combustion Chamber Refractory

Water contact will also disturb the adhesives that hold the refractory in place, reducing their strength and adding to the likelihood of refractory displacement.

A different problem occurs once the appliance is fired. The moisture and trapped water in the refractory will heat and turn to steam, subjecting the refractory to the expansion process and leading to spalling and refractory degradation. The moisture can be minimized by frequent, shorter firing periods in an effort to slowly dry and drive out the moisture. A visual assessment of the refractory during this process is needed.

Rodent Damage

During their off-season, boilers or furnaces that do not operate year-round are attractive locations for rodents. It is common for rodents looking for food, nesting material, or nesting places to damage the refractory components of the appliances (Figure 10-39). Technicians should anticipate the damage and address the problem during servicing.

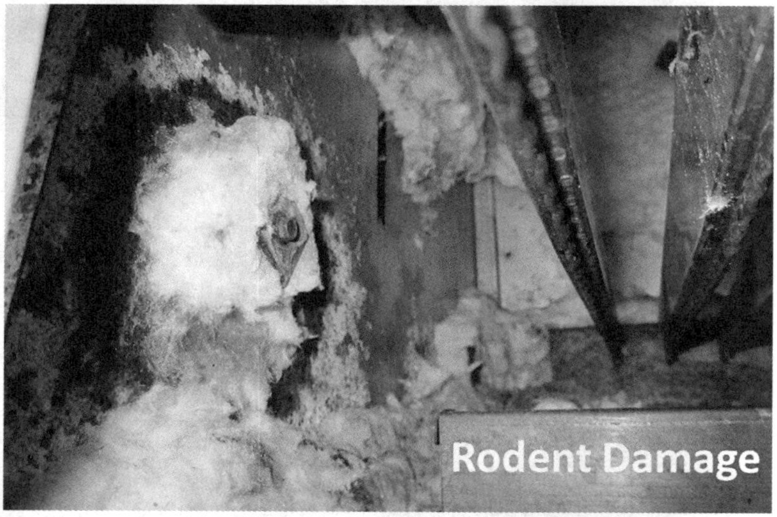

FIGURE 10-39 Rodents Have Damaged This Boiler's Combustion Chamber Refractory During Off-Season Use

Traps or rodent repellents may be successful in preventing or stemming the damage. However, take care never to leave these in the appliance's combustion chamber unless steps are taken to prevent the appliance from accidentally firing.

Effects of Failed Combustion Chamber Refractory

A failed combustion chamber refractory can cause soot to discharge into homes (Figure 10-40) and can also cause fires. Maintaining this component in top condition is vital to ensure that an appliance will operate safely for the upcoming heating season, prolonging its life expectancy and preventing damage to the home.

FIGURE 10-40 Soot Damage to a Home

Refractory Obstruction of Flame

If pieces of a refractory fall from their intended location in the appliance's combustion chamber, the space where the burner flame is supposed to be positioned will now be "shared" with the flame and fallen refractory piece. This is true regardless as to if the appliance is oil or gas fired. This causes the flame to strike the fallen pieces, resulting in flame distortion and soot production (Figure 10-41).

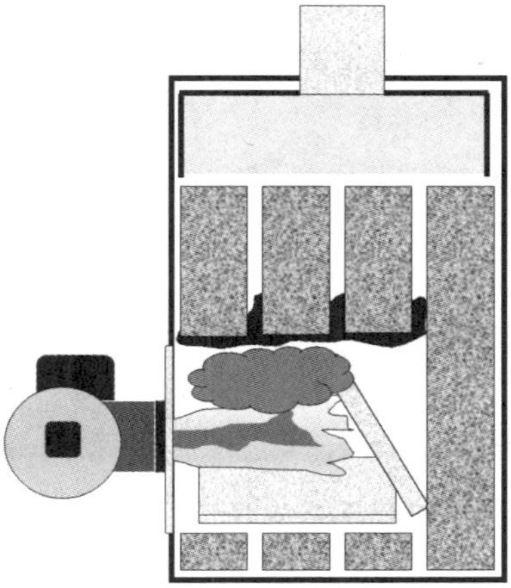

FIGURE 10-41 Oil Burner Flame Impingement Caused by a Fallen Combustion Chamber Refractory Piece

If the level of degradation of the refractory is severe enough, the entire refractory component may collapse (Figure 10-42), accelerating soot production and leaving large areas of the furnace or boiler firebox exposed to the high heat of combustion.

FIGURE 10-42 Collapsed Refractory and Flame Impingement Hole in Firebox

In gas-fired boilers or furnaces, pieces of a failing combustion chamber refractory will fall onto the gas burner's flame spreader and block the flow of gas (Figure 10-43). Left unchecked, disruption in the combustion process occurs, resulting in soot production (Figure 10-44).

FIGURE 10-43 Refractory Debris Has Fallen on to Gas Burners

FIGURE 10-44 Soot Production Has Obstructed Boiler Flue Passageways

When an appliance suffering this kind of flue obstruction is operated for a prolonged period of time, the soot will eventually plug the furnace or boiler and discharge into the structure. This can occur in both oil- and gas-fired appliances (Figure 10-45).

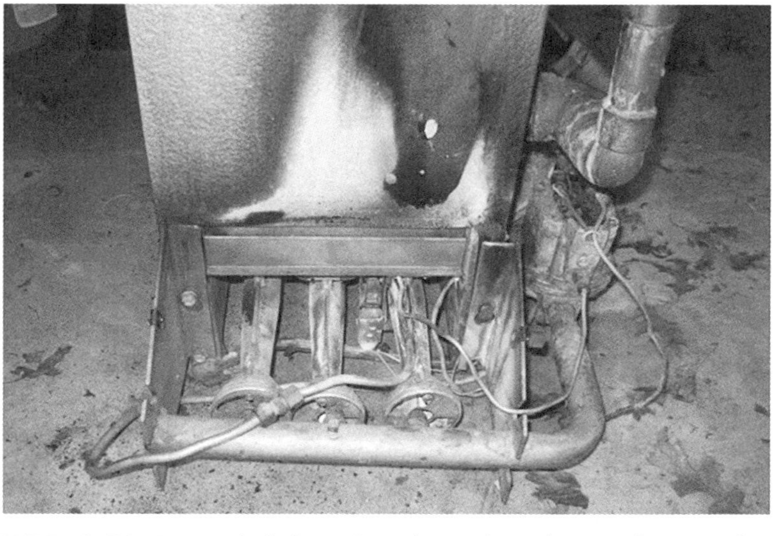

FIGURE 10-45 Reversal of Flow of Products of Combustion from Appliance

Missing Refractory and Flame Impingement

A failed refractory will allow the burner flame to contact the steel components of a forced hot air heat exchanger or a boiler's combustion chamber components. Because the temperature of the burner flame exceeds the temperature of the melting point of the steel, it causes the heat exchanger to fail because of *flame impingement*.

A hole in an oil-fired heat exchanger (Figure 10-46) is especially dangerous since there is an increased likelihood of products of combustion passing from the primary or combustion side of the heat exchanger to the secondary side. This results in products of combustion causing damage to the home (also known as a puffback) and physical health risks.

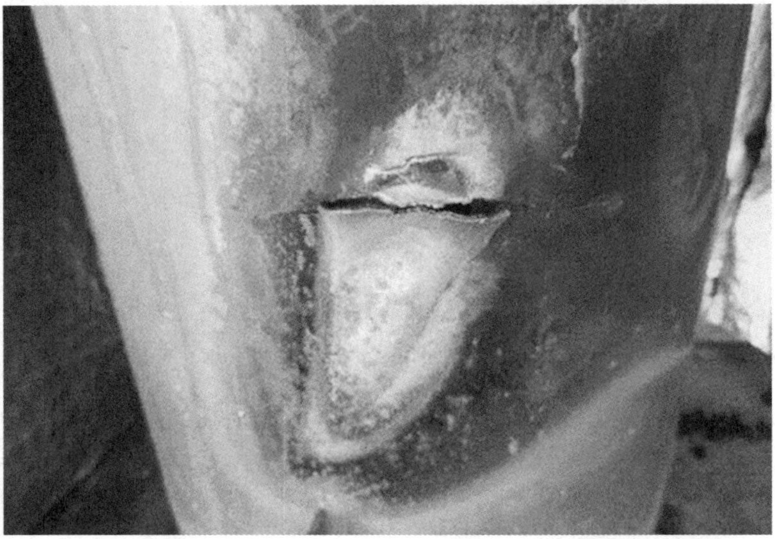

FIGURE 10-46 Flame Impingement Hole in Furnace's Heat Exchanger

Missing refractory components will also result in damage to dry base boilers of both oil and gas combustion. Damaged and missing refractory panels in a dry base boiler (Figure 10-47) allow high-heat damage to the steel firebox portions of the appliance.

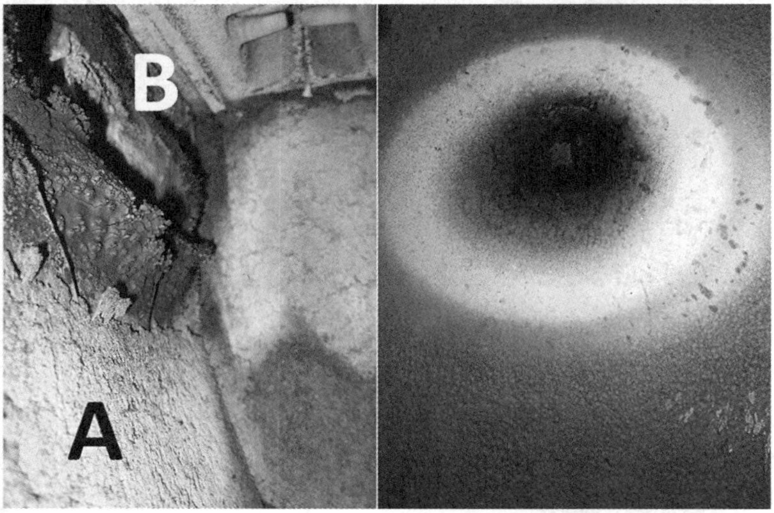

FIGURE 10-47 (A) Damaged Refractory Panels in Gas-Fired Boiler Allowed Damage to (B) Firebox Construction

Repair of the Refractory Versus Replacement of the Appliance

The service technician's responsibility when performing preventative maintenance on a boiler or furnace is to exercise professional judgment in determining if the appliance can operate for the foreseeable future in a safe manner. Assessment of a combustion chamber's refractory is part of this preventative maintenance and is an action that a service technician should be expected to accomplish with reasonable accuracy. And remember that oftentimes, assessing a combustion chamber's refractory means having to decide whether to replace the appliance rather than repair it by just replacing the chamber.

In making the decision, consider that refractory kits and components are relatively inexpensive in cost and labor. In most residential appliances, the job can take from 1 to 5 hours, depending on the design of the appliance. When this is considered against the cost of replacing a boiler or furnace, refractory replacement is a worthwhile maintenance action—as long as there are no other problems with the appliance.

Also, before deciding to repair a failing combustion chamber, rather than replace it, make sure that the replacement parts are available and the repair action is within budget.

In replacing the combustion chamber refractory, keep these points in mind:

- The integrity of the combustion chamber refractory is vital to the safe operation of a boiler or furnace. Since inhalation of products of combustion can cause injury or even death, accurate assessment of this component is an important preventative maintenance action during yearly service events.
- If an appliance with a combustion chamber that needs to be replaced is allowed to run until something actually breaks (aka reactive maintenance), this can result in premature failure and unnecessary replacement of the entire appliance. Additionally, if the burner flame strikes the fallen refractory, the flame will not burn as intended and produce soot. This soot obstructs the appliances flue passageways and allows combustion gas leakage into the home. Additionally the exposed firebox or heat exchanger in the appliance will eventually result in a flame impingement hole preventing the hot combustion gasses from being contained within the appliance.
- Always consider the visual indicators of wear and tear of the refractory. Always err on the side of safety. And if you are in doubt, err on the side of caution and be proactive in your maintenance actions.
- Finally, remember, always handle refractory materials in a safe way and in accordance with manufacturer's recommendations. Ceramic fiber refractories can require that you wear breathing and skin irritation protection whenever you are removing old material, vacuuming the firesides of product-of-combustion debris, or installing new chamber refractory material!

CHAPTER 11

BOILER WATER CHEMISTRY

With newer low-volume heat exchangers as well as increasingly more diverse numbers of metals and alloys used in their construction, more attention to boiler water chemistry will be required to assure proper operation as well as adherence to manufacturers' warranty requirements.

Furthermore, heating system impurities such as sludge and corrosion by-products will inhibit water flow through the system, will accelerate corrosion damage to the appliance's heat exchanger, and may allow localized overheating of boiler heat exchanger components.

Corrosion damage to a boiler reduces its life-span, and sludge and scale accumulations in the heat exchanger reduce the rate of heat transfer. The result is that the boiler will need more fuel than it should, thus increasing operating costs to the homeowner.

Annual preventative maintenance actions should include testing the boiler water and adding chemicals if needed. By doing so, you are providing the homeowner with a service that will allow the maximum life from the boiler and assure that the efficiency of the system is as intended by the manufacturer. Ignorance isn't bliss, and remember that especially in older systems, boiler longevity and system efficiency are dependent on good flow and heat transfer; as such, to properly maintain the appliance, your attention to the system water is vital.

The first step in maintaining good and manageable boiler water quality, even for small residential boilers, is to set up a program of water quality monitoring and maintenance actions to correct any identified issues.

Hydronic and Steam Heating Systems

As stated in Chapter 2, heat is transferred from the boiler to the rooms in the building by either hot water (Figure 11-1) or steam (Figure 11-2).

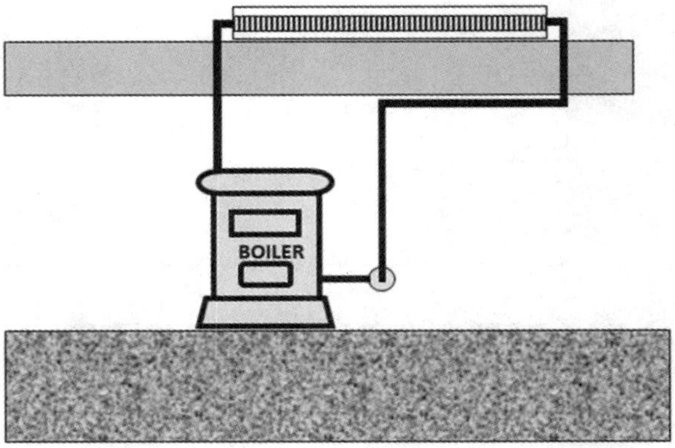

FIGURE 11-1 Hot Water Boiler System

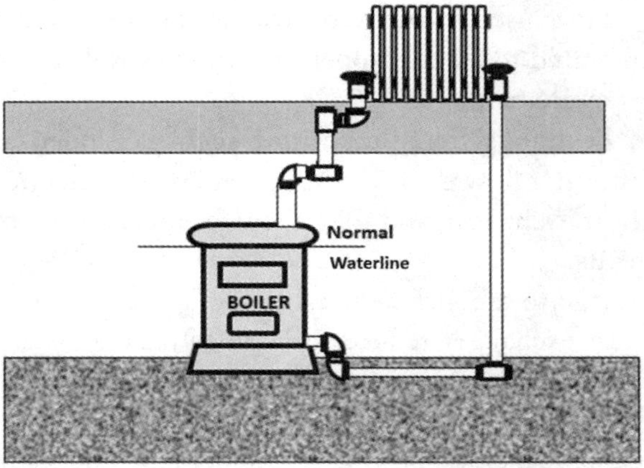

FIGURE 11-2 Steam Boiler System

Ideally, these are supposed to be "closed systems," meaning that the water that is initially added to the boiler and system piping should remain in the system without any losses. In reality, however, this is not true even for forced hot water systems. The loss of water creates water chemistry issues that are detrimental to the operation and service life of the boiler supplying hot water or steam to the system.

All boilers have the potential to leak a small amount of water. This may be from push nipples or gaskets between the sections, and this leaking water may be identified through maintenance inspections. Regardless, when the boiler is in operation, often water will leak into the boiler firesides and evaporate, and the leak will never be known to anyone not focused on this possibility. This is true for both steam and water boilers.

Steam boilers, however, are not true closed systems since in addition to losing water if there is a boiler leak, water and steam are commonly lost through air vents discharging steam along with air. This requires the addition of makeup water, which in turn makes the system an "open" system rather than a truly closed one. The addition of makeup water creates an ongoing cycle where the boiler water chemistry is changing and usually with detrimental effects.

Cycles of Concentration

Initially, a water or steam boiler and heating system piping are filled with water of a certain chemistry that contains a fixed amount of dissolved hardness minerals. You can liken these dissolved minerals to dissolved grains of sugar in a cup of coffee. As noted earlier, if the boiler or heating system components lose water in the form of steam, the loss of water is countered by the addition of makeup water. However, as the water or steam is lost from the steam heating system, the dissolved minerals remain in the system water, causing these dissolved solids to accumulate (Figure 11-3).

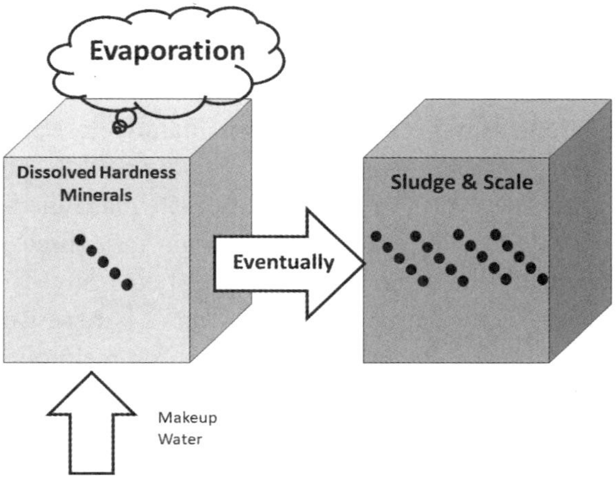

FIGURE 11-3 Cycle of Concentration

As an example, say the source water contains 300 parts per million (ppm) of these total dissolved solids (TDSs). Due to the continual leakage and makeup cycle, the heating system's accumulated TDSs may increase to 900 ppm. This is equal to *3 cycles of concentration* for the heating system. Given the boiler water's particular chemistry, 3 cycles of concentration may be acceptable allowing the system to operate without these hardness minerals (that are still in a liquid form) turning back into a solid and forming sludge or scale, however, as the cycles of concentration increase, so does the likelihood of the boiler water chemistry forming these undesirable deposits.

Using a conductivity meter, you can identify the amount of dissolved hardness minerals in a sample of water as well as the number of cycles of concentration of these hardness minerals and make corrections if needed.

Cycles of concentration = conductivity of system water/
conductivity of makeup water

Different areas of the world also have different hardness minerals to contend with. For example, silica and chlorides are two minerals often found in water. To calculate their cycles of concentration, you would use these equations:

Cycles of concentration = silica in system water/silica in makeup water

Cycles of concentration = chlorides in system water/chlorides in makeup water

If no evaporation has taken place, the number of cycles of concentration will be 1.

Without adding chemicals to the boiler water to change the concentration of the minerals, much in the way chemicals are added to commercial boilers, the ability of a boiler to operate at elevated cycles of concentration may be difficult to maintain. As such, it is good practice to observe and document the correlation between water clarity and sludge formation in relation to the cycles of concentration over a period of time to determine the best level of cycles of concentration for the particular system.

One way to reduce the cycles of concentration is to drain and then refill the boiler. Other options may be available from manufacturers of water treatment chemicals.

What happens as the cycles of concentration increase in the heating system? Let's use our cup of coffee analogy. Much like adding two, three, four, and then five spoons of sugar, eventually the sugar can no longer be dissolved in the limited volume of coffee. Likewise, high cycles of concentration of hardness minerals in a boiler system will allow these previously dissolved minerals to precipitate out of solution to form sludge and scale.

Sludge and scale accumulation in a boiler is detrimental to boiler operation. For one thing, it changes the boiler water chemistry. For another, it acts as an insulator, decreasing boiler heat transfer efficiency and occasionally contributing to overheating damage. Sludge and scale also interfere with the operation of boiler controls such as low-water cutoff switches, aquastats, and safety relief valves.

Corrosion By-Products

How old is the house that has the boiler you're servicing?

Steam boilers were an 1800s technology that went out of favor for residential heating in the early 1900s. The fact is, the people that installed the heating system in a house that is heated with steam are long gone. The boiler may have been replaced, but the steam supply, condensate return, and radiators (Figure 11-4) are still there, distributing steam for heat and returning condensate today.

Gravity and then forced hot water systems—the early successors to steam systems—are also still in use in older homes. Just as in a steam system, the boiler may have been replaced. However, the same original piping is still in place and functioning, as seen in the photographs of the home built in the 1920s depicted in Figure 11-5.

FIGURE 11-4 Steam Heating System Piping and Radiators Still in Use

FIGURE 11-5 Older Forced Hot Water System from House Built in the 1920s

Corrosion is a large problem with these older systems. Not only do corrosion by-products enter the system from aging piping, but corrosion from the watersides of the boiler itself creates a condition where these solids are forming and concentrating in the boiler water and settling in the system. This is evident in the side-by-side comparison of the water-level gauge glasses of two different steam boilers, as depicted in Figure 11-6. The photo on the left of the figure shows a well-maintained boiler with good water quality; the photo on the right shows the effects of concentrations of corrosion and minerals in the boiler water.

FIGURE 11-6 Water Clarity in Steam System

Sludge and Scale Removal

While sludge and scale can form in either water or steam boiler systems, these deposits are obviously more prevalent in steam systems.

Initially when these substances form, they accumulate in lesser amounts. At this point, they mix with boiler water and have a viscous semiliquid form. This is sludge.

Given the additional cycle-of-concentration solid mineral buildup as well as system corrosion by-products, sludge turns into hard scale (Figure 11-7). Scale is harder to remove than sludge. To reduce scale formation, the proactive and easier maintenance action is to remove the sludge from the system before it has a chance to turn into scale.

FIGURE 11-7 Scale Deposit on Heating System Piping Inner Surfaces

Always refer to the boiler manufacturer's specific instructions regarding this and all other maintenance actions. However, in general, testing for and removing sludge from the system is done by first shutting off the electricity to the boiler and letting the water cool for a period of time. This reduces the possibility of burns and scalding and will prevent thermal shock from damaging the boiler when cold makeup water is added quickly.

After the boiler water has cooled sufficiently, drain off a sample, say between 1 and 2 gallons. Do this at the boiler drain valve (Figure 11-8) or at the lowest portion of the system. Pay attention to the consistency and color of what is being drained.

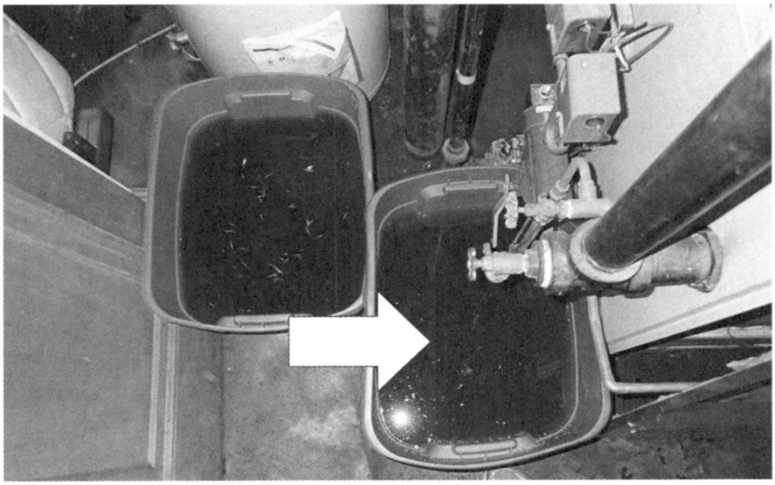

FIGURE 11-8 Draining Boiler Water

If the color and consistency of the water sample indicates that sludge and scale are present or if poor system flow or past piping repairs reveal an unacceptable buildup of scale and sludge, use one of the commercial boiler water treatment products that are available to get rid of the unwanted substances. This is generally done by adding a mixture of boiler water and boiler chemical into the system and letting it circulate for a period of time before the system is emptied and refilled. Continually drain and refill the boiler, flushing it until the water quality has improved and a visual examination shows little sign of entrained solid particles (Figure 11-9).

FIGURE 11-9 Boiler Water Clarity Comparison

Steam Boiler Cleaning

New boilers with castings contain a large amount of oil and other chemicals. As water is added and the boiler is fired off, these oils more readily mix with the boiler system water. Additionally, when new boilers are integrated with steam boiler system piping that likely was installed a century or so ago, the mere act of manipulating the boiler piping knocks off sludge and scale that was otherwise solidified in place. These substances too are introduced into the new boiler's water content, resulting in a cocktail of boiler water that has an undesirable water chemistry.

Draining, refilling, and, if need be, flushing and draining the boiler again can, to a great degree, remove dislodged sludge, scale, and oils from the system. Make sure, though, that before you do so, the boiler has cooled to a safe degree, and *do not ever allow the boiler to fire in a low-water condition.* Also, adding cold water to a hot boiler can cause damage to the boiler's heat exchanger.

Foaming

Impurities in the boiler water result in foam generation, which interferes with steam flow from the boiler and also causes carryover where liquid from the foaming enters the steam pipes. This causes the water level in the boiler to rapidly rise and then fall (also known as *hunting*) (Figure 11-10).

At the point where the water level is high, water is carried over onto the steam side of the system, which in turn causes knocking and water-hammer banging in the steam pipes as the steam travels through the system. At the point where the water level is low, the boiler's low-water cutoff switch shuts the burner down, resulting in frequent starting and stopping of the burner.

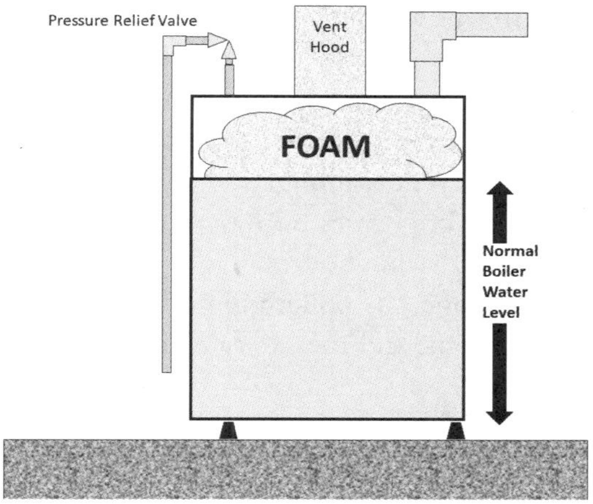

FIGURE 11-10 Foam Generation in a Steam Boiler

Skimming

The impurities in new boiler systems can be removed by a cleaning process known as skimming. The new boiler is fired and operated for an initial "break-in" period during which time the impurities are skimmed from the boiler pressure vessel. Skimming is done by creating a bleed-off pathway from the top of the boiler. The pressure relief valve is replaced with a tee fitting to which the boiler's safety relief valve is connected as well as a temporary globe valve. This temporary globe valve is connected to discharge piping to throttle the discharge from the boiler to control the skimming process.

If the boiler is outfitted with an opening or fitting connection closer to the normal operating condition's waterline, this is a better place to attach the skimming piping since it eliminates the need to operate the boiler at a higher than normal water level. When the boiler is fired and made operational, the water level is raised higher than normal just to the top of the boiler pressure vessel (Figure 11-11).

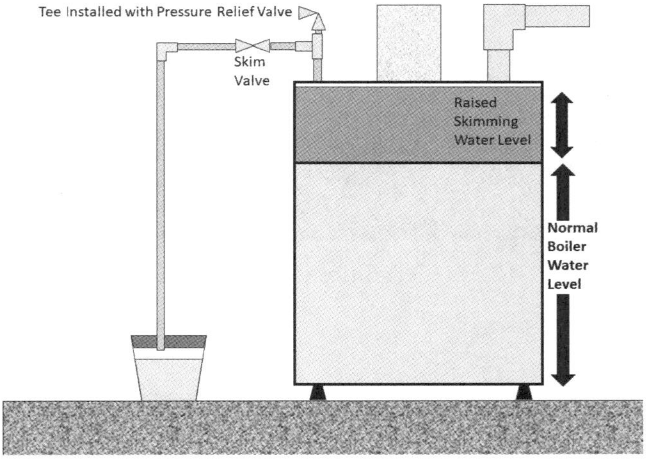

FIGURE 11-11 Removing Impurities from a Steam Boiler aka "Skimming"

By firing the boiler in this arrangement and keeping the skim valve just barely open, impurities that will rise to the top of the boiler water can be removed over a number of days of operation. Care must be exercised to remove the collection of sediments and contaminants from the skim discharge bucket, and the effluent from the boiler should be covered to prevent scalding.

In addition to skimming impurities from the surface of boiler water, you can use chemicals made specifically for steam boilers.

When the cleaning is done, the boiler piping should be reconfigured to match the manufacturer's specifications, and the safety relief valve should be properly connected to the boiler pressure vessel.

pH

The alkalinity or acidity of a solution is known as its pH, which is a measure of the solution's hydrogen ion concentration. A pH value less than 7 is acidic, and a value greater than 7 is alkaline (a pH of 7 is neutral). The pH value of a solution affects the other parameters of water chemistry. pH can be controlled by the addition of commercially available water treatment chemicals.

After the boiler is cleared of sludge and a clean water sample is obtained from the boiler's drain valve or riser service drain, pH test strips (Figure 11-12) or a pH meter can easily be used to identify the boiler's pH level. If the pH does not match the boiler manufacturer's recommendation, a pH-adjusting and corrosion-inhibiting treatment chemical can be added to the boiler.

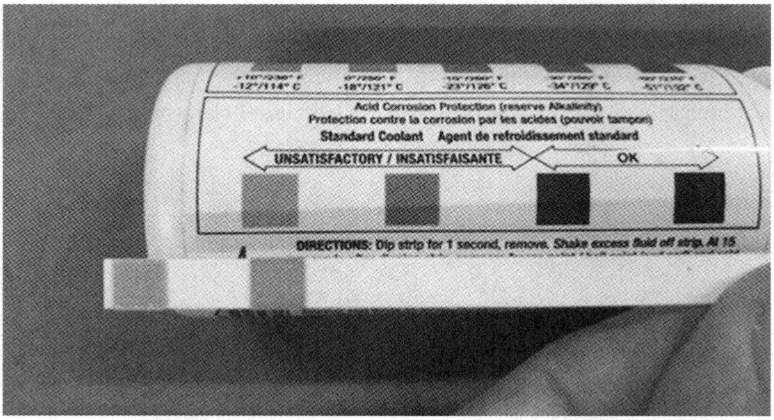

FIGURE 11-12 pH Test Strips (Courtesy of Oatey Products)

MAINTENANCE TIP!

Different Boilers, Different Water Treatment Requirement

Don't be "the guy" that has to tell a homeowner that a leak in the owner's 4-year-old wall-mounted condensing boiler was caused by corrosive water and that it is not covered under the manufacturer's warranty because you didn't know that you needed to test and adjust the pH of the boiler water and add an antifreeze–corrosion inhibitor chemical. *You* recommended the boiler; *you* installed it; *you* were asked to maintain it; *you* should know what you're doing!

Different Boiler Materials Require Different Water Chemistries

Boiler water chemicals and pH levels may differ among manufacturers, and so too will heat exchanger construction and material. For example, heat exchangers may be made from aluminum (Figure 11-13), stainless steel (Figure 11-14), steel, or the more familiar cast iron (Figure 11-15). The different materials used in the construction of the various boiler heat exchangers require different boiler water chemistries, and as such, it is a necessity to verify that the chemicals used in cleaning and in system operation are compatible with the appliance and system. Long-term boiler water additives like antifreeze or other recommended chemicals likewise should be verified by the manufacturer for heat exchanger compatibility.

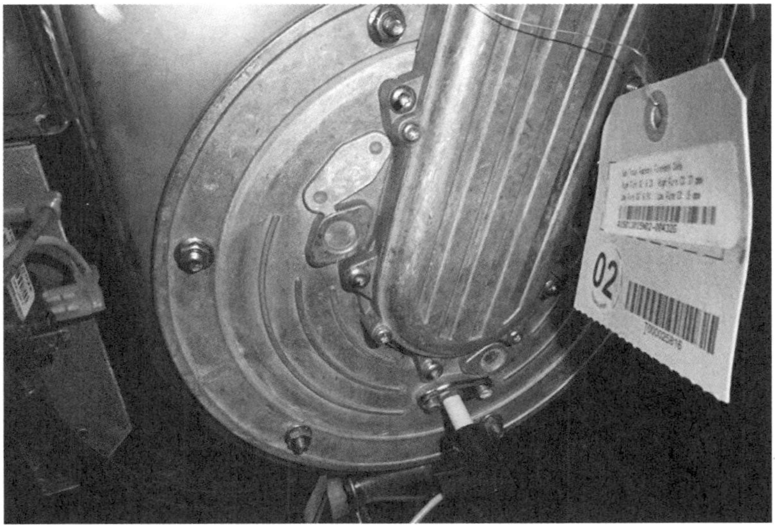

FIGURE 11-13 Aluminum Boiler Heat Exchanger

FIGURE 11-14 Stainless Steel Heat Exchanger

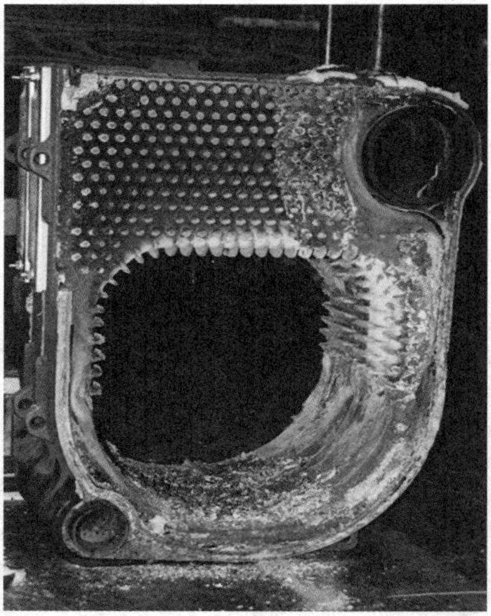

FIGURE 11-15 Cast-Iron Heat Exchanger

Water chemistry tests of pH and other parameters are required annually to satisfy the requirements of many boiler warranties regarding boiler heat exchangers. Failure to perform and present verifiable documentation of these tests may void manufacturers' warranties for boiler heat exchangers that have failed due to defects or corrosion-related failures. Also, ensuring the correct water chemistry ensures heat exchanger longevity and efficiency of the system.

Oxygen and Other Dissolved Gases

When makeup water is added to a heating system, oxygen and other noncondensable gases in the water are also added to the system. To reduce as much as possible the amount of makeup water that will be needed, it is important that leaks and losses from forced hot water and steam boilers be repaired. Prolonged exposure to highly oxygenated and corrosive boiler water chemistries results in corrosion failures and increased levels of iron oxide in systems and boilers.

Forced hot water boilers are less prone to damage from oxygen and other dissolved gases since these boilers usually do not operate in an extreme condition of a fully open system. Also, they do not increase the cycles of concentration of solids greatly since they usually do not have exorbitant amounts of oxygen-rich boiler makeup water added to them.

Steam boilers, however, do have the tendency to release water (in the form of steam) from their systems, which in turn requires the addition of makeup water, which in turn adds oxygen and other noncondensable gases on an ongoing basis. This creates a corrosive environment on the waterside of the boiler above the steam boiler's waterline and manifests itself as oxygen corrosion damage, including that in cast-iron steam boilers as seen in Figure 11-16.

FIGURE 11-16 Oxidation Corrosion Hole in Boiler Section

The loss of steam and hence boiler water, which in turn causes excessive makeup water to be added to the boiler, is assessed by investigating leaks during preventative maintenance actions.

Some problem areas causing the loss of water from steam systems are difficult to find. For example, buried condensate drain piping under the basement floor can be an unknown source of a leak. Corrosion holes or other damage in the hidden piping can result in the loss of water from the system. Infrared imaging cameras may be used to identify hot water or condensate leaking from piping that otherwise would have remained hidden from view (Figure 11-17).

FIGURE 11-17 Infrared Image Showing Leaking Hot Water Pipe Under Concrete Floor

Leaking steam vents in steam boilers are another source of water loss (Figure 11-18). Annual maintenance of these systems should include a visual and operational examination of the vents to make sure that they are not constantly leaking, causing steam and water loss from the system.

FIGURE 11-18 Steam Header Steam Vent

Antifreeze

Some systems have piping in areas of the building prone to freezing temperatures. For example, the hydronic piping of hydro air systems with air handlers might be installed in unheated attics (Figure 11-19) or in unheated building cavities such as crawlspaces (Figure 11-20). Antifreeze is used in these systems to prevent damage to the hydronic hot water piping.

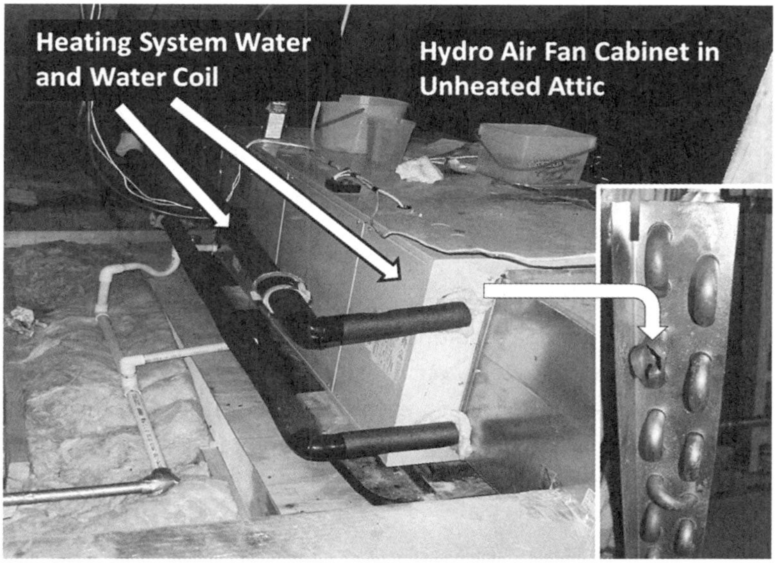

FIGURE 11-19 Hydronic Water Piping Supplying an Unheated Attic-Installed Hydro Air Handler Must Be Protected Against Freezing

FIGURE 11-20 Hydronic System Piping in Unheated Crawlspace

There is also the possibility that during bitterly cold and unexpected weather, a power outage may occur. If this happens, freeze-protected system piping in unheated areas can buy time to prevent water damage due to freezing until electric power is restored and the heating system can again become operational.

Be sure to use inhibited propylene glycol antifreeze since it is formulated specifically for hydronic systems. Other antifreeze solutions intended for other types of systems such as automotive systems should not be used.

Both the American Society of Heating Refrigeration and Air Conditioning Engineers (ASHRAE) and manufacturers recommend that antifreeze levels be monitored and tested on a regular basis to ensure that the amount of antifreeze is sufficient and that the system is protected against freezing temperatures expected in the upcoming winter.

There are several ways to check antifreeze levels:

One way is to use a hydrometer, which is a syringe-like water extractor that contains different-colored pellets. Each of the pellets reacts to the water's specific gravity, resulting in a positioning of the pellets within the hydrometer that indicates a particular solution strength.

Another device to measure antifreeze strength is a refractometer. A refractometer (Figure 11-21) utilizes the varying light transmission properties through antifreeze-protected solutions to identify the freeze protection ability of the antifreeze.

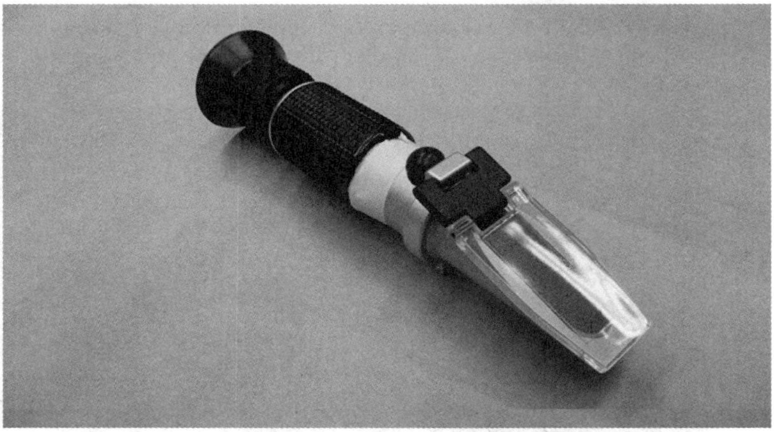

FIGURE 11-21 Water Testing Refractometer

Still another way to verify antifreeze protection levels is to use test strips. After an acceptable clear water sample is obtained, a test strip is dipped in the sample water for 1 second (Figure 11-22), removed, and allowed to react for a 15-second period. A color reference card or table is used (Figure 11-23) to compare the results of the test strip with known concentrations to identify actual water conditions.

FIGURE 11-22 Testing a Sample of Boiler Water

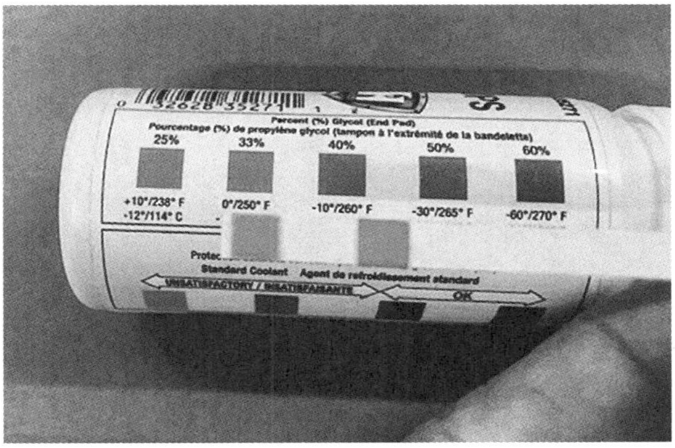

FIGURE 11-23 Antifreeze Test Strips (Courtesy of Oatey Products)

Weak levels of antifreeze can become corrosive and damage system piping and appliances. To combat this problem, most companies offer heating system antifreeze products that also provide a corrosion inhibitor.

Adding Water Treatment Chemicals

Steam Boilers

Once the boiler water is free of sludge and scale, small amounts of boiler additive chemicals can be added directly to a boiler. However, the chemicals should be diluted and administered according to the manufacturer's instructions to prevent damage or too strong a chemical concentration. To gain access to the boiler (once it has cooled down and has no pressure on it), remove the safety relief valve or other threaded boiler pressure vessel opening. Then add the chemicals directly into the boiler pressure vessel.

Water Boilers

Water boilers typically do not have the water chemistry issues that steam boilers do. However, it is still very important to install and maintain them with proper chemistry in accordance with the manufacturer's recommendations.

The water in the boiler will require annual testing and potential adjustment to maintain proper levels of pH and antifreeze and also to ensure that corrosion is not a problem. For example, initial tests for clarity and impurities (Figure 11-24) may indicate that a corrosion inhibitor needs to be added to the system.

FIGURE 11-24 Sampling Boiler Water from Drain Valve

When adding antifreeze or other water treatment chemicals to a water boiler supplying a forced hot water system, the additive must be circulated throughout the entire system. Adding chemicals to the boiler alone allows additives to be diluted once they eventually mix with system water. This may result in insufficient strengths of antifreeze and other needed additives.

Chemicals are added to forced hot water boilers as well as those supplying hydro air hydronic heat pumps with the use of a small auxiliary "pony pump." This is a small rubber impeller pump outfitted with hose connections and is available at plumbing supply warehouses. It is suggested that a pump in the 6- to 10-gallon-per-minute range be used.

Many boilers were initially plumbed with the intent of facilitating the addition of boiler water chemicals to the system water. As such, access and shutoff valves may not have been installed, and this leaves the technician with the task of having to invent a way to add chemicals or retrofit the heating system piping with the needed fittings.

As the technician, how might you go about it? A starting point is knowing that the boiler drain and service valves are outfitted with hose bib connections, which are convenient connection points for chemical injection pumps.

First, turn off the electricity to the boiler and wait until the system water has cooled to a safe level (about 90°F). Then draw off a sample from the bottom drain to determine if the system water is free of sludge or unwanted impurities. If these are found, draining and refilling may be needed until the water quality is acceptable.

Connect the discharge side of the pony pump to the boiler's outlet (heating supply) side, allowing water from the pony pump to be pumped into the system piping (Figure 11-25).

FIGURE 11-25 Connecting Pony Pump to Boiler Outlet Service Valve

Place the inlet (suction) side hose connection to the pony pump in a 5- to 10-gallon bucket or pail, and connect the inlet to the pail to the lowest point of the boiler's heat exchanger, the bottom drain valve (Figure 11-26).

FIGURE 11-26 Connection to Lowest Part of Boiler's Heat Exchanger

Check the valve alignment, including zone valves, making sure they are manually placed in the open position.

Add needed antifreeze or other water treatment additive (and a small delivery amount of water) to the bucket. Follow the manufacturer's recommendations about the concentration of antifreeze and other additives regarding strength and cold temperature protection. The pony pump transfers the water from the bucket to and through the system piping. Regulate the water level in the bucket by opening or closing (throttling) the boiler drain valve so that the flow rate of water into the bucket equals the flow rate going out, making sure the pump does not run dry nor the bucket overflow.

Move the pony pump discharge hose to other zones to ensure that all heating system zone systems are treated and that the net chemical concentrations of the entire system (boiler and all zones) are of adequate strength. This results in a flow circuit through the boiler's heat exchanger, the various zones of hydronic water protecting the boiler, and all system piping, ensuring consistent chemical mixing (Figure 11-27).

After the water is thoroughly circulated through all zone systems, check the antifreeze or other water treatment chemical level for proper strength. After you are satisfied that the system is properly treated, reconfigure the system piping, including the opening of the water feeder valve, and adjust the boiler water pressure to the proper operating pressure (between 11 and 13 psi or according to the manufacturer's recommendations).

The burner can then be started and the system monitored for operation. During this time, the cold, newly treated boiler water will heat and expand, testing the function of the expansion tank. During this start-up period, any deficiencies in the expansion tank may become noticeable. Also, if the expansion tank is too small, failed, or overwhelmed by the heating and expansion of this initial large volume of water, the boiler's safety relief valve may lift.

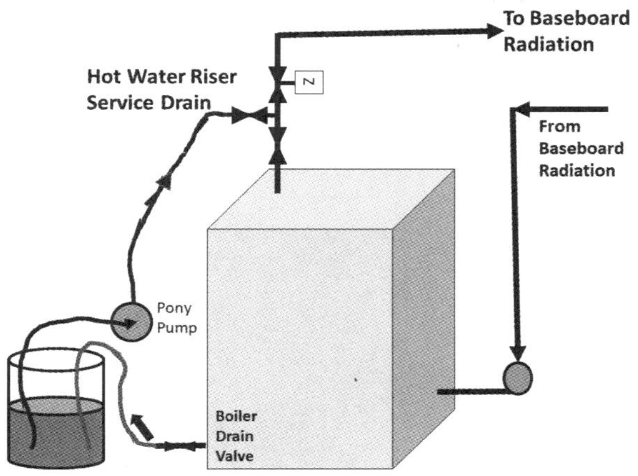

FIGURE 11-27 Chemical Addition for Forced Hot Water System

It is important to make sure that the system achieves thermal equilibrium after the water treatment chemicals have been added and the burner has been off for a period of time since this will test the system's ability to function in an elevated pressure condition. Also, make sure the system is outfitted with a properly operating backflow preventer since boiler chemicals cannot be allowed to flow back and enter the domestic water supply!

References

ACCA Manual J Residential Load Calculation.

Flynn, D., *The Nalco Water Handbook* (3rd ed.), McGraw-Hill, New York, 2009.

Hydronics Institute Guide H-22.

I=B=R Guide to Residential Hydronic Heating.

IECC.

IMC.

Oatey Inc., "4 Tips to Extend the Life of a Boiler."

CHAPTER 12

BOILER AND FURNACE ANNUAL INSPECTION AND CLEANING REVIEW

This chapter is a review of some of the important points that were made earlier in the book. It begins with this reminder: *The time to make repairs to equipment is when the deficient condition is small, the damage is minor, and the repair is easy.* Equipment does not "heal" itself, and time is your enemy since a small issue will most certainly grow into a larger one.

Most manufacturers of heating equipment of all fuel and energy sources recommend an annual maintenance event. The maintenance event may be making repairs to damaged items (reactive maintenance), addressing expected maintenance requirements such as replacing filters and performing cleaning (preventative maintenance), or inspecting and making a visual assessment of key components in the appliance to ensure that they can survive the next operating period until the next annual or planned maintenance event (predictive maintenance).

The homeowner is relying on *you* and *your* expertise for advice about what maintenance has to be done to keep the boiler, furnace, or heating system operating until the next *planned* maintenance event.

Unless you have an endless supply of customers who will simply allow you to install new appliances whenever a repair requiring any type of effort is needed, you will soon run out of work and develop a reputation not favorable to your bottom line. The annual or planned maintenance event is your opportunity to catch problems before they develop into larger ones and to avoid unplanned emergency repairs that are more stressful and expensive to perform for both you and the homeowner.

Remember, your expertise is not limited to a particular component of the appliance or system. Oil burner technicians, for example, are not limited to servicing the burner alone. The entire appliance is your domain. This includes everything from the burner, appliance controls, and exhaust and fuel systems, to any interfacing or add-on items such as humidification and air-conditioning components.

If, for example, you are servicing a blower fan on a furnace that has failed and you notice that an air-conditioning evaporator coil has been backing up and allowing a furnace to be damaged by water contact, *say something*! *Do something* such as telling the homeowner and getting permission to make repairs or if need be replace the component or appliance.

In essence, your customers' heating and conditioning equipment is *your* equipment. Yes, they own it, but it will be you who has to fix it. Keep your house in order and stay on top of needed repairs while they are small and not overwhelming.

And be sure to write things down. We live in an age where our phones are cameras and everyone is connected to the internet 24/7/365. Develop a system with your office where observed deficiencies and needed repairs are photographed, explained, and verifiably *communicated* to your customer. In a simple letter, e-mail, or phone call to a customer, state what was found, what is needed, and what the consequences of avoiding the repair are. Also, diligently written service records kept in the customer file as well as posted on the appliance detailing findings and recommendations will help another technician assigned to the job later on as well as prove that you made proper recommendations.

Finally, if the appliance is not safe to operate and could cause damage to the home or occupants and if the homeowner does not show a willingness to make the needed repairs, it's better to condemn the appliance and shut it down rather than let it continue on its path to catastrophe.

In addition to the manufacturer's suggested maintenance items, the following are areas of focus in the annual and periodic maintenance event.

Cleaning

Although cleaning the firesides of a boiler's heat exchanger is typically associated with servicing an oil-fired appliance, the heat exchangers in gas-fired boilers, including low-mass condensing boilers, must also be inspected as well as cleaned if needed.

Before beginning, it is a good idea to initiate a combustion performance test. For oil-fired equipment, start with a smoke test. If the appliance passes this (with a zero to trace of smoke), the other tests of the combustion performance test can be performed. This serves as a baseline for comparison after the appliance is cleaned and returned to service.

Cleaning an appliance will generate soot, ceramic fiber refractory particles, and other by-products of combustion. Breathing and skin protection is needed.

In preparation for cleaning the heat exchanger of a boiler or furnace, it is necessary to remove the draft hood, open the cleanout plates, and remove the burner. This also exposes as much as possible the surfaces of the heat exchanger that are in close proximity to the burner flame. This provides the opportunity to visually assess damage in an area where heat damage is most likely and that would otherwise go unnoticed.

To clean the heat exchanger, use a vacuum designed specifically to remove and collect the fine particles of products of combustion (Figure 12-1). Take care to strategically place the vacuum hose into one area of the appliance while the other is being cleaned. This allows any disturbed dust to be drawn into the vacuum and collected without it being discharged into the air adjacent to the appliance (Figure 12-2).

FIGURE 12-1 Soot Vacuum (Courtesy of Mastercraft)

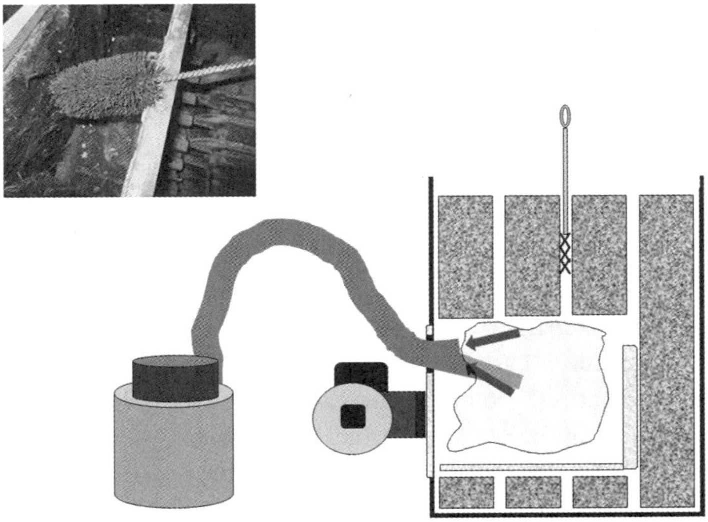

FIGURE 12-2 As the flue passageways are brushed, escaping soot can be contained and prevented from spreading by using a vacuum placed in an area where dislodged particles can be collected before they spread into the home.

Thoroughly removing any accumulated scale and soot is one of the most important things you can do to assure an efficiently operating combustion appliance. If left on the heat exchanger surfaces, scale and soot will act as an insulator and reduce heat transfer thermal efficiency.

Combustion Chamber Refractory Assessment

As mentioned in Chapter 10, visual assessment of the combustion chamber refractory (if installed) is most easily accomplished during the cleaning process, making note of any thinning, cracking, and curling, all indicators of refractory deterioration. Also as mentioned in Chapter 10, care must be taken to assure that there is no damage to the refractory caused by cleaning brushes striking it during cleaning (Figure 12-3).

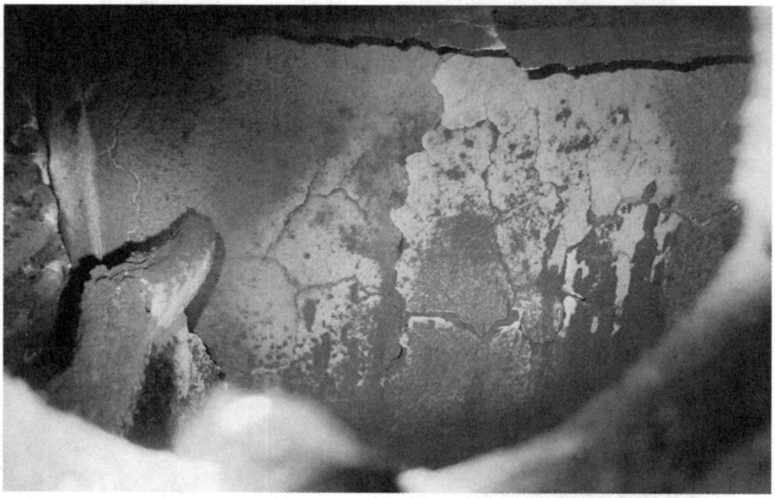

FIGURE 12-3 Assessment of Combustion Refractory Shows Impending Failure

Inspection for Boiler Water Leaks

With the boiler partially disassembled for the cleaning process, now is a good time to examine the heat exchanger for leaks. Focus on common areas for leaks such as the joining push nipples or gaskets between cast-iron boiler sections as well as piping connections to the heating system (Figure 12-4).

Remember, a steam boiler runs partially filled with water with the upper third of it occupied by steam. As per Chapter 11, oxidation corrosion holes can develop, and depending on where they are situated, they may not be easily identifiable (Figure 12-5).

The best time to look for holes and leaks is at the start of a heating season. Begin by overfilling the boiler to the point where the water completely fills the boiler (and no more than that). Any preexisting holes or leaks will become evident, allowing you to find them when there is time to make repairs or replace the boiler before cold weather sets in.

FIGURE 12-4 Boiler Connections Are Leaking as Seen During Annual Inspection

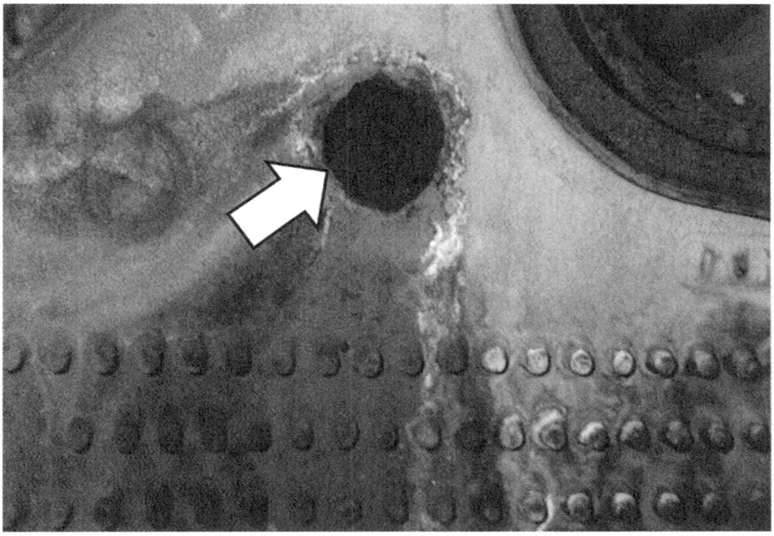

FIGURE 12-5 Oxidation Corrosion Hole Above Waterline

Tankless hot water coil gaskets as well as gasket connections on circulator pumps should be examined for signs of wear and leakage.

Initially, a gasket will display a brittleness in the form of gasket cracking. This will likely preclude leakage to the gasket fit watertight joint.

Pay special attention to the tankless hot water coil gasket. If this often-overlooked boiler component is leaking and not fixed, corrosion will progress to the point that the boiler will need to be replaced before the end of its expected life-span (Figure 12-6).

FIGURE 12-6 Leaking Tankless Hot Water Heat Exchanger Gasket

Keep in mind that the older a tankless hot water coil gasket is, the more likely it will be a problem. Any joint leakage should be caught and corrected as early as possible since leakage can cause not only corrosion but also seizure of the connecting bolts, which makes replacement of the tankless coil difficult if not impossible.

Tankless hot water coil replacement (Figure 12-7) can extend the life of a boiler. Note that the replacement should be scheduled during planned (nonemergency) maintenance in the off-peak season.

FIGURE 12-7 Replacing Tankless Hot Water Coil Heat Exchanger and Gasket

Control Maintenance

Boiler Controls—Steam

Water-connected controls such as low-water cutoff switches have the potential of leaking, and as such, they should be examined to address this possibility. Water that comes into contact with control mechanisms may allow stray current electrical flow, high-resistance heat-generating electrical connections, and seizure of mechanical components, causing control failure and the resulting damage (Figure 12-8).

FIGURE 12-8 Leaking Low Water Cutoff Switch Must Be Replaced (Courtesy of McDonnell & Miller–Xylem Inc.)

Boiler Controls—Forced Hot Water

On water boilers, controls susceptible to leakage and resulting corrosion damage include pressure-reducing (water feeder) valves (Figure 12-9). A valve that is leaking is indicative of an internal synthetic rubber diaphragm failure. The leak may manifest itself as water leaking from the joined gasket connection or from the valve stem adjustment.

Air-eliminating valves and vents are also an area of potential water leakage. This problem should be anticipated and evaluated during preventative maintenance events. Internal valve seats are prone to fouling caused by hardness minerals and other debris, preventing the seat from fully closing. This causes continual leakage that not only adds to the vent failure but damages adjacent boiler auxiliary components (Figure 12-10).

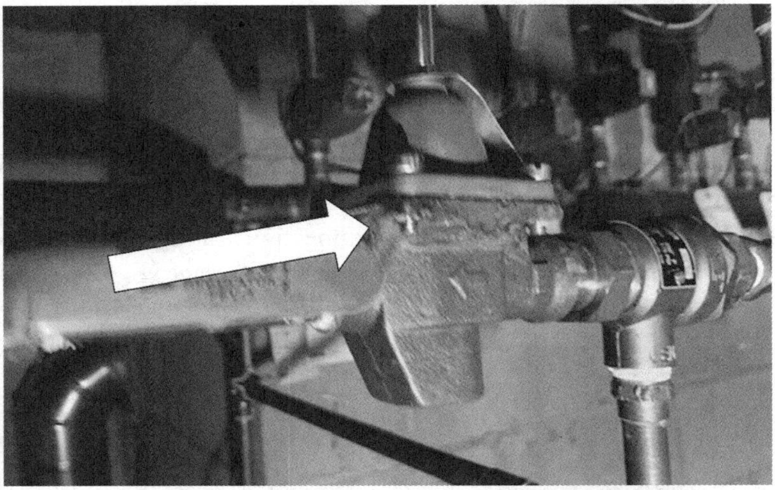

FIGURE 12-9 Leaking Diaphragm on Water Feeder Valves (Courtesy of Watts Industries)

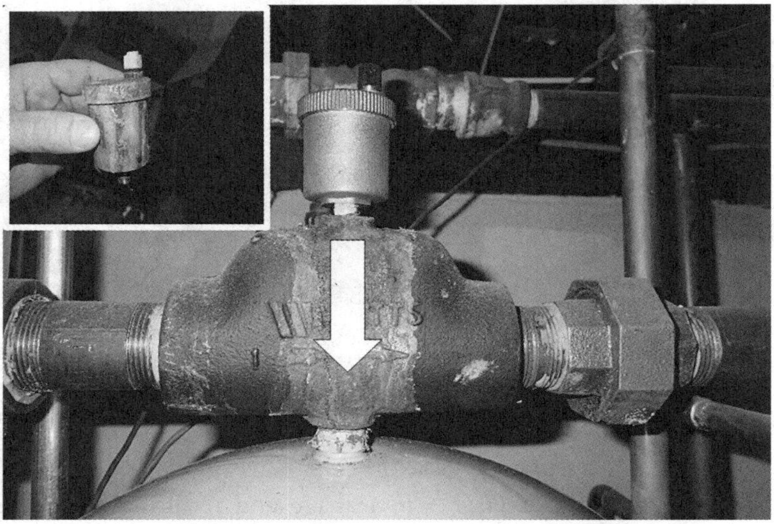

FIGURE 12-10 Leaking Air Vent

Expansion Tank

Forced hot water system expansion tanks should be tested for a proper charge of the air side of the tank as well as any leakage signs.

The use of a tire pressure gauge can identify if sufficient air pressure consistent with the manufacturer's recommendations is present. Also, any water leakage from the air admission valve is an indication that the internal synthetic rubber bladder has ruptured. A rupture will cause the air side of the tank to become "waterlogged" and prevent the expansion tank from being able to compensate for pressure increases in the system that result from the heating and cooling of system water (Figure 12-11).

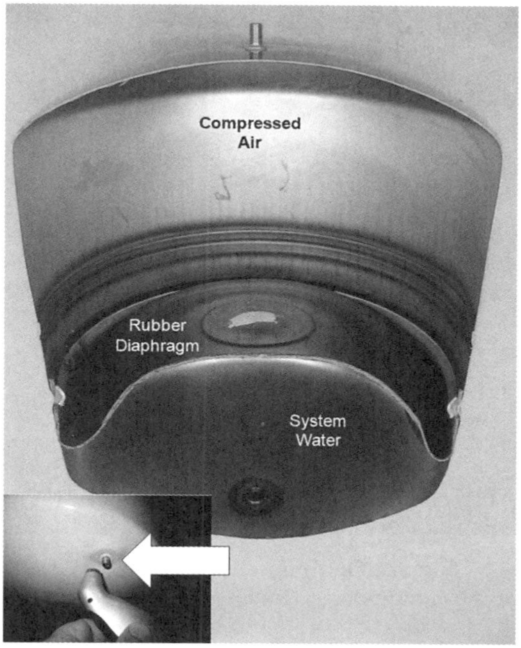

FIGURE 12-11 Bisected Expansion Tank

Furnace Water Leaks

Leaking or Overflowing Air-Conditioning Evaporator Coil Condensate Pan

Many gas- or oil-fired forced hot air furnaces also serve summer air-conditioning needs with the installation of a refrigeration evaporator coil in the plenum, adjacent to the furnace (Figure 12-12).

Proper collection and diversion of the generated air-conditioning condensate away from the furnace is vital to prevent this condensate water from damaging the furnace and controls. Generated condensate on the evaporator coil falls into and is collected in the air-conditioning evaporator condensate pan (Figure 12-13), where it is then plumbed to a suitable drain.

FIGURE 12-12 Furnace Outfitted With Evaporator Coil Allowing It to Provide Air Conditioning Cooling During Summer Months

FIGURE 12-13 Condensate Collection Pan Under Air Conditioning Evaporator Coil

As well, standing water, hardness minerals, and entrained dust particles will, over time, concentrate and allow the growth of microorganisms. These impede water flow and can allow the condensate collection pan located just under the air-conditioning coil to overflow. Ongoing water leakage due to an overflowing air-conditioning condensate pan causes damage to the furnace's heat exchanger and also to other components and controls, as seen in Figure 12-14.

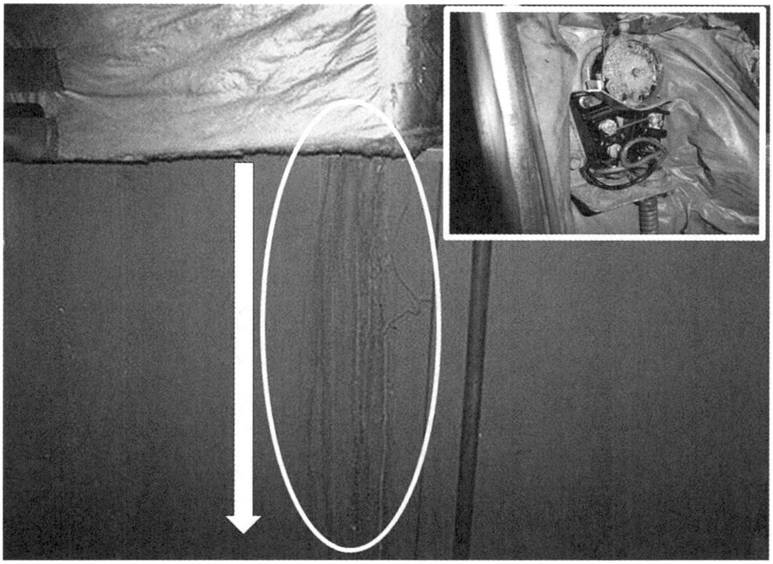

FIGURE 12-14 Damage to Furnace Caused by Air Conditioning Evaporator Coil Condensate Pan Overflowing

Evaporator Coil Maintenance

Signs of condensate leakage should not be ignored. Rather, they should be addressed immediately. Affected components and controls should be inspected and replaced if needed.

Careful attention to furnace heat exchangers exposed to water from any source including an overflowing evaporator coil condensate pan should be thoroughly inspected for leaks and progressive corrosion, and if the damage is deemed excessive, they should be replaced. Additionally, water accumulation in the air handler or furnace cabinet has the potential to contribute to mold growth, which can be harmful to the occupants in the home rendering this component of the system in need of replacement as well.

Preventative maintenance of the components that make up the condenser—both the parts that produce the condensate and the parts that remove it—should be performed prior to or after the summer "cooling season" before a problem develops. In those areas of the world where air conditioning is needed year-round, schedule a preventative maintenance call at least on an annual basis or more frequently if identified conditions require it. In any case, preventative maintenance should be performed even if it is a mere visual evaluation of existing conditions.

Preventative maintenance of the parts of an air-conditioning system that remove condensates should include the following.

Evaporator Coil Cleaning

Opening the evaporator coil housing or ductwork plenum will provide access to the copper tube–aluminum-finned evaporator coil refrigerant heat exchanger. If dust has accumulated on the surfaces of this component, the dust can be removed by light vacuuming and the use of commercially available and manufacturer-approved cleaning agents (Figure 12-15).

FIGURE 12-15 Inspecting Evaporator Coil

Evaporator Condensate Pan Cleaning

A large portion of the contaminants that cause fouling of the condensate removal system are in the water-collecting condensate pan. Accumulated debris in this area can be removed by the use of a wet vacuum as well as commercially available cleaning chemicals. Some technicians have been reported to use simple white vinegar as a cleaning agent. However, again manufacturer's recommendations should be followed to ensure that warranty obligations are met.

Condensate Drain Line and/or Condensate Pump Cleaning and Wet Vacuuming

Just as debris that has accumulated in the evaporator pan stops the flow of water and causes a backup, debris that has accumulated in the drain piping can cause clogs in a drain or condensate removal pump (Figure 12-16). Clearing accumulated water and debris can be accomplished with a wet vacuum, which will pull standing nonflowing water from the system. Additionally, coil cleaners or other commercially approved cleaning agents listed for this purpose can be used.

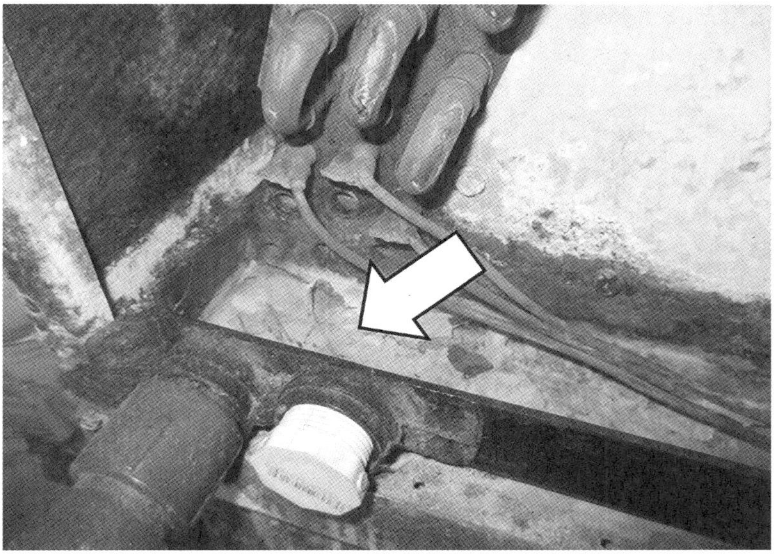

FIGURE 12-16 Debris Collection in Evaporator Coil Condensate Collection Pan

Condensing Gas-Fired Furnace

In gas-fired high-efficiency condensing boilers and furnaces, the combustion process is so efficient that products of combustion are reduced in temperature to below the dew point of the discharge combustion gases. This causes entrained water vapor to condense into water.

In the combustion process, various draw points of condensate removal are present throughout the appliance. Common locations are the induction fan located at the end of the combustion process (Figure 12-17) and the product-of-combustion discharge connection to the vent at the outlet of the furnace's burner system.

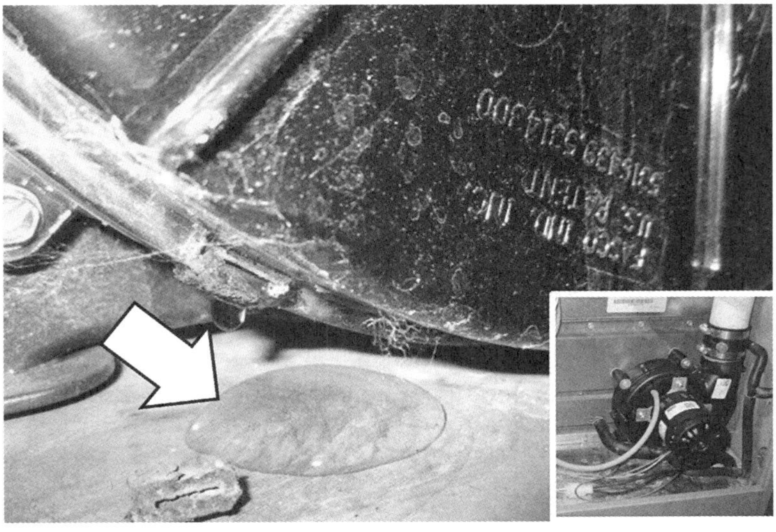

FIGURE 12-17 Condensate Leakage from Furnace Combustion Fan

Water connections to the induction fan as well as the fan housing itself may show signs of water leakage of condensate, which must be addressed. Not only can prolonged water contact cause corrosion damage to the furnace and result in mold formation in the cabinet. There is also the possibility that water will leak into the electrical controls, causing appliance shutdown, malfunction, and electrically induced fire damage to the appliance (Figure 12-18).

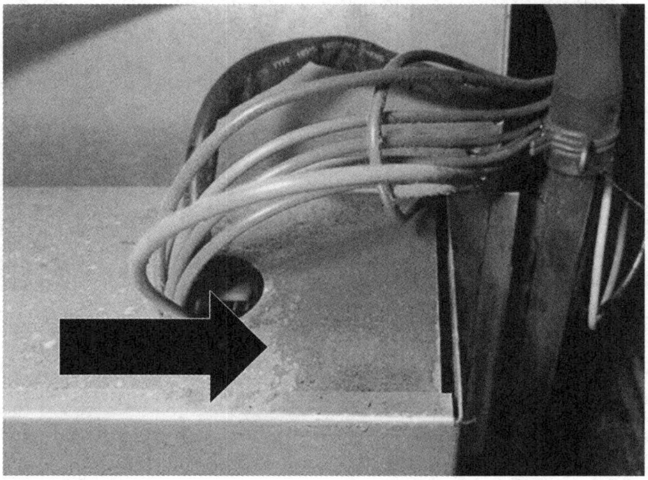

FIGURE 12-18 Condensate Leakage onto Furnace's Control Cabinet

Boiler Water Chemistry

Boiler water can be tested quickly and easily. The idea is to identify problems and needed adjustments to combat the detrimental effects of poor boiler water chemistry. Unwanted heat exchanger corrosion, system sludge, and scale formation, as well as frozen components, can all be eliminated or greatly reduced by adjusting the boiler water chemistry to the levels required by the manufacturer.

TDS and Cycles of Concentration

Cycles of concentration are a measurement of the amount of total dissolved solids in boiler water and the ratio of these solids to the amount that was in the water initially. This measurement is accomplished with a conductivity meter (Figure 12-19). The higher the cycle of concentration, the more likely dissolved minerals (silica, chlorides) will precipitate out of solution and become a solid. These solids form sludge and scale. This problem is more likely to occur in steam boiler systems than in forced hot water or hydro air hydronic systems since water boilers should not be losing water at a rate that would increase the cycles of concentration greatly.

In systems where water loss may be an issue it is suggested that the cycles-of-concentration be measured, observed, and documented over a period of time to determine if they are increasing. If so, this is an indication that there are leaks from the system that may be in need of repair. Other options may be available from manufacturers of water treatment chemicals as well.

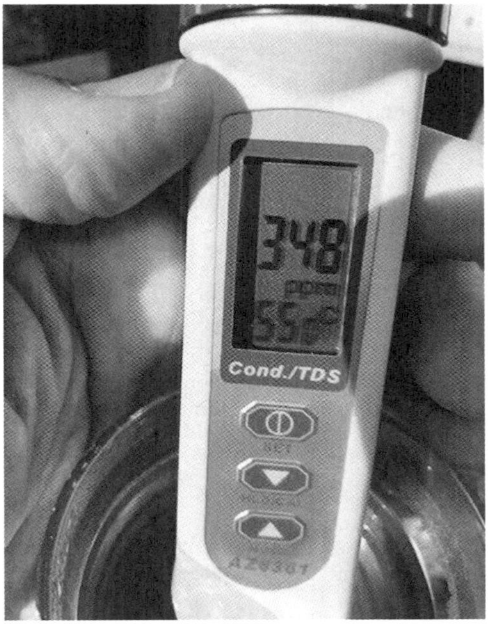

FIGURE 12-19 Conductivity Meter

Corrosion By-Products

Visually examining the boiler water is another way to determine what is occurring in the system. Opening the bottom boiler drain and sampling a quantity of boiler water may reveal sludge and other sediments in the system that may be indicative of an unnoticed leak or corrosive boiler water chemistry (Figure 12-20).

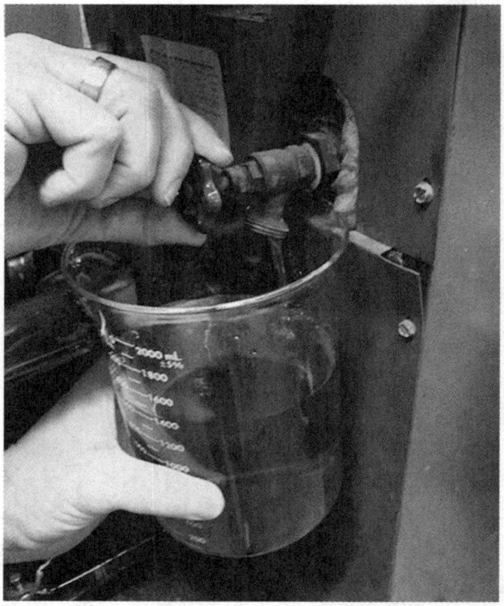

FIGURE 12-20 Collecting and Visually Examining Boiler Water for Contaminates Within the System

pH

pH is a measure of the acidic or alkaline condition of the boiler water. The pH scale ranges from 0 to 14, where 7 is neutral, below 7 is acidic, and above 7 is alkaline. pH can be measured using test strips or a pH meter or by doing chemical testing. Keep in mind that for many different boiler designs and materials, different boiler water chemistries are not only recommended but required by manufacturers in order to maintain the manufacturers' warranties.

Antifreeze Levels

Some heating system pipes in forced hot water or hydro air systems are located in vulnerable areas that are prone to freezing such as exterior abutting walls or unheated attic spaces. If the weather turns extremely cold, it can cause power outages or excessive thermostat setback, and the heating pipes may freeze (Figure 12-21). In these systems, the addition and maintenance of antifreeze plays a crucial part in keeping the heating system operating; and like any other portion of the system, the pipes must be maintained on a regular basis in accordance with the manufacturer's requirements. As well, antifreeze levels must be maintained in accordance with recommendations from organizations such as ASHRAE.

FIGURE 12-21 Freeze Damage to Forced Hot Water Baseboard Piping

Combustion Gas Leaks

Once the boiler or furnace is cleaned and returned to operation, you need to verify that the heat exchanger is able to contain products of combustion without leaking.

For oil-fired appliances the over-the-fire draft measurement will indicate if there is a possible obstruction in the appliance or chimney flue passageways or venting fan that will result in combustion gas leaks.

An ambient air measurement of carbon monoxide or other products of combustion is an indicator of product-of-combustion "spillage" that requires attention. This test can also be done on the space between the heat exchanger and jacket to determine if combustion gas leaks are present (Figure 12-22).

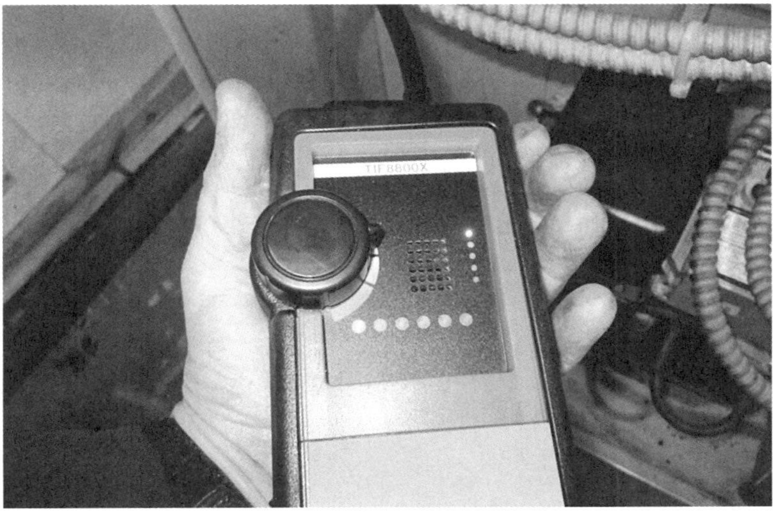

FIGURE 12-22 Product-of-Combustion Gas Detector (Courtesy of Bosch–Robinair)

Also CO or other products of combustion identified in an analysis of the supply air from a forced hot air furnace may be an indication of a leaking cleanout door or a heat exchanger failure. Combustion gas leakage can also be an indication of combustion side gasket failure, requiring that the door gasket be replaced (Figure 12-23).

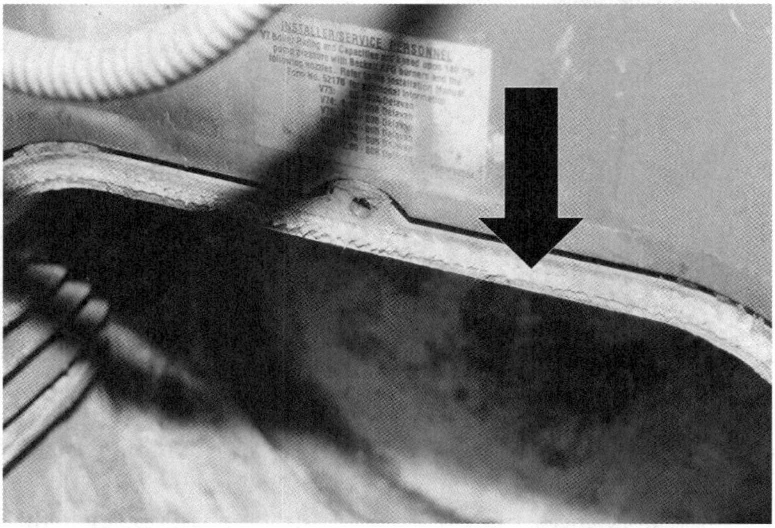

FIGURE 12-23 Examining Boiler's Combustion Chamber Door Gasket

The Combustion Performance Test

You cannot identify the efficiency or quality of combustion of a boiler or furnace by eye alone. A combustion performance test, composed of a number of diagnostic tests, should be performed to detect unacceptable or dangerous conditions that cannot be detected visually.

It is crucial that you document the results of your combustion performance–efficiency test for your records. Also, be sure to post them on the appliance in the event that *you* are not the next technician to service the appliance or if you forget what you did the last time you were there. In addition, leaving a record of what you did shows that you did all within your power during the last service event to make sure that the appliance was operating correctly (Figure 12-24).

Chapters 6 and 7 provide more detail about how combustion performance tests are conducted. The following is a review of the needed tests that together make up the entire combustion performance evaluation of the overall efficiency of the appliance.

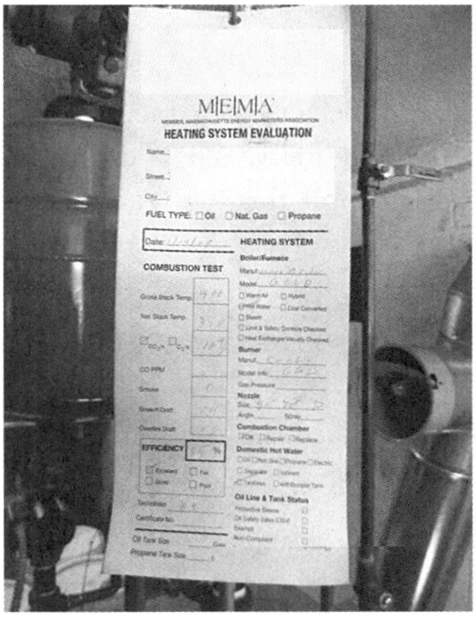

FIGURE 12-24 Combustion Performance Test Slip Left on Appliance for Homeowner and any Following Technicians Review

Ambient CO Test

A product-of-combustion discharge from the appliance indicates an obstructed furnace or boiler flue passageway. If the type of combustion chamber is not a sealed combustion chamber with air piped to and exhaust gases piped from the appliance, combustion spillage will result. In Figure 12-25, for example, an improperly firing gas burner produced soot that obstructed the heat exchanger "cell," causing product-of-combustion backup and spillage into the space where the appliance was installed.

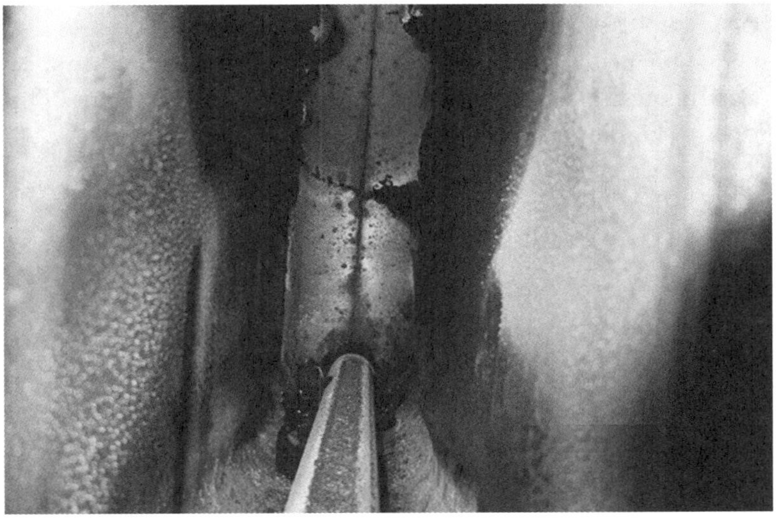

FIGURE 12-25 Examining Firesides of Gas-Fired Furnace's Heat Exchanger

Smoke Test

Performed on oil-fired appliances, the smoke test determines if complete combustion is occurring. Smoke levels should be zero to a trace of smoke, and this level of adjustment should be achieved first (Figure 12-26). Performing the other elements of a combustion performance test on appliances with an unacceptable smoke reading may damage these instruments.

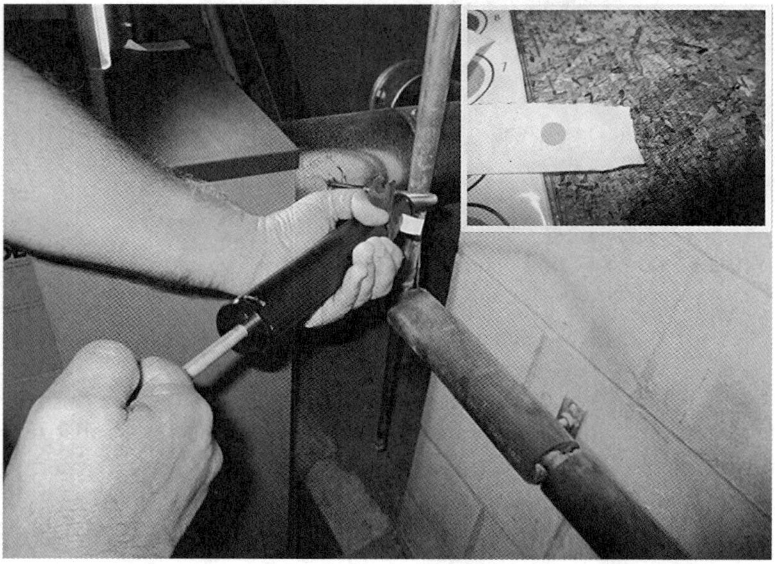

FIGURE 12-26 Performing Smoke Test

Testing for Over-the-Fire Draft–Oil

In an oil-fired appliance, a manometer or draft gauge can be used to measure over-the-fire draft. This test is used to determine if the heat exchanger is obstructed or drafting properly (Figure 12-27).

FIGURE 12-27 Measuring Over-the-Fire Draft with Electronic Manometer

Testing for Breach Draft

The test for breach draft verifies the measurement of gas flow from the appliance to and through the chimney or exhaust fan. Low draft may be an indication of a chimney obstruction, or it may mean that a draft inducer booster fan is required (Figure 12-28).

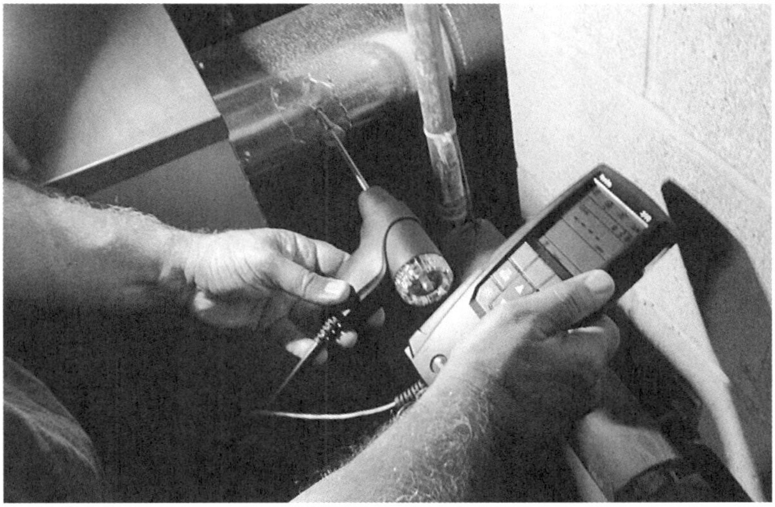

FIGURE 12-28 Measuring Breach Draft

Measuring Oxygen or Carbon Dioxide Content

Oxygen (O_2) or carbon dioxide (CO_2) content is measured with either a computerized electronic combustion analyzer or a conventional fyrite-based instrument (Figure 12-29). The CO_2 or O_2 levels in the flue gas indicate the efficiency of the combustion occurring in the appliance.

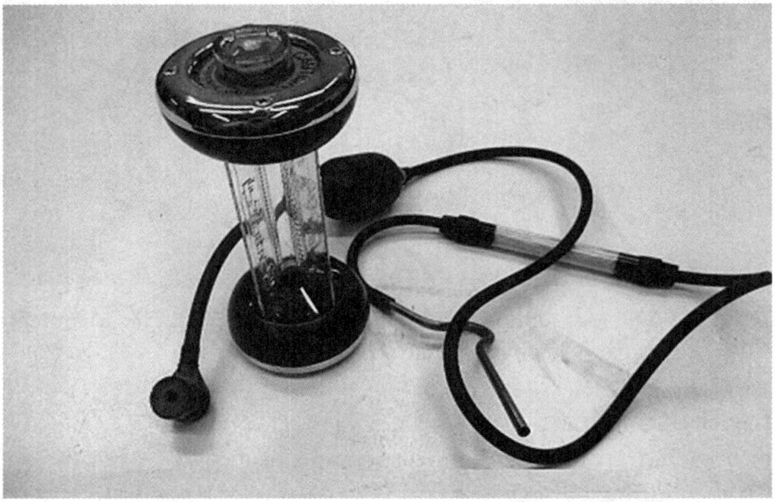

FIGURE 12-29 Fyrite Type of Combustion Analyzer

INDEX

About the Author

JOHN CERTUSE is a registered professional mechanical engineer licensed throughout the northeast, as well as a licensed oil burner technician, pipe fitter, and certified fire and explosion investigator.

John has served as President of the National Academy of Forensic Engineers (NAFE) and is a graduate from the Massachusetts Maritime Academy with a degree in Marine Engineering. John served as a Lieutenant in the Navy Reserve and owns patents for fuel usage analysis as well as for freeze protection piping insulation designs.

John is the owner of ISE Engineering in Attleboro Massachusetts, providing fire and forensic investigation services throughout the northeast.